Power Systems

D.P. Buse and Q.H. Wu

IP Network-based Multi-agent Systems for Industrial Automation

Information Management, Condition Monitoring and Control of Power Systems

 Springer

D.P. Buse, PhD
Q.H. Wu, PhD, CEng, FIEE

Department of Electrical Engineering and Electronics
The University of Liverpool
Brownlow Hill
Liverpool L69 3GJ
UK

British Library Cataloguing in Publication Data
Buse, David P.
  IP network-based multi-agent systems for industrial
  automation : information management, condition monitoring
  and control of power systems. - (Power systems)
  1. Electric substations - Automatic control - Computer
  network resources 2. Intelligent agents (Computer software)
  3. Electric power transmission - Data processing
  4. Distributed artificial intelligence
  I. Title II. Wu, Qing-Hua
  621.3'126'02854678
ISBN-13: 9781846286469

Library of Congress Control Number: 2006939313

Power Systems Series ISSN 1612-1287
ISBN 978-1-84628-646-9   e-ISBN 978-1-84628-647-6   Printed on acid-free paper

© Springer-Verlag London Limited 2007

9 8 7 6 5 4 3 2 1

Springer Science+Business Media
springer.com

To our parents and families

Preface

Due to the complexity of distributed systems, such as railways, aerospace systems, navigation systems, gas transmission systems, power utility and power plants, the conventional automation system is not capable of providing information management and high-level intelligent approaches. This is because achieving these functionalities requires comprehensive information management support and coordination between system devices, and the control of many different types of task, such as data transportation, data display, data retrieval, information interpretation, control signals and commands, documentation sorting and database searching *etc.* These operate at different timescales and are widely distributed over the global system and its subsystems. Without reasonably designed system software architectures and hardware structures, it is impossible to handle these tasks efficiently, safely and reliably, with the possibility of online reconfiguration and flexibly embedding applications.

With advances in communication technologies, in particular Internet technology, in recent years, Internet Protocol (IP) networks have been considered for use in conventional automation systems. This book is substantively concerned with developing a novel concept−e-Automation−that capitalises on the advantages of IP networks and agent technology for system integration and leads to a new generation of industrial automation systems. In contrast to conventional automation systems, an e-Automation system can provide integrated functionalities for distributed information management condition monitoring and control with an open architecture of system software and hardware for implementation of various tasks within Wide Area Networks (WAN) and Local Area Networks (LAN). The e-Automation system is able to provide great gridability and communication capability to resolve the problems of task implementation and information management for a wide range of distributed complex industrial systems.

The basic idea of the multi-agent-based e-Automation system was first considered in 1998, based on work undertaken in the Intelligence Engineering and Automation research group, The University of Liverpool, over the previous ten years, in the areas of distributed control and automation systems, computational intelligence, intelligent systems, information management, and power system control and operation. Since then, e-Automation architecture has evolved with initial support from National Grid Transco (NGT). As the e-Automation system is in its early stage of development, some fundamental issues need to be studied. What is the optimal hardware structure of the system? What is the most suitable software platform that can accommodate the intelligent agents and give room for future development? How can one develop a stable, reliable, and robust system? Is there any theory to support the system design? Many unsolved issues need to be investigated.

This book describes a substation automation architecture based on the concept of e-Automation, using the multi-agent systems methodology. The book begins by presenting the historical background of substation automation systems, along with the newer network-based approaches and architectures. Agents, multi-agent systems and mobile agents are also introduced. The main contributions of the book are concerned with the development of an agent-based architecture and its components for power system automation. The implementation of a substation information management system with multi-agent-based architecture is also presented.

We would like to thank John Fitch and Zac Richardson of NGT, for supporting this work and providing assistance with the substation simulator and Information Management Unit (IMU), and Brian Baker, also of NGT, for his support of the work during its initial stages. We would also like to thank Pu Sun, who worked on the hardware architecture and the data acquisition system for the prototype, and with whom we collaborated on the writing of several papers based on the work, and Jun Qiu Feng, who worked on the human−machine interface and personal agents for the prototype system, which are included as part of Chapter 6 in this book. Finally, thanks go to Chen Ma for a large amount of time spent on the preparation of diagrams and typesetting as well as the provision of the standards of agent development platforms.

This work was supported by NGT. The facilities of the Department of Electrical Engineering and Electronics and the e-Automation Laboratory at the University of Liverpool were essential to the completion of the work.

Special thanks go to Anthony Doyle (the Senior Editor), Sorina Moosdorf (the Production Editor) and Simon Rees (Editorial Assistant) for their professional and efficient editorial work on this book. Our thanks are also extended to all colleagues in the Intelligence Engineering and Automation research group, The University of Liverpool, for all assistance provided, and which have not been specifically mentioned above.

University of Liverpool, UK,   *David P. Buse*
                                                                     Qing-Hua Wu

August, 2006

Contents

List of Figures

List of Tables

1

Introduction

A conventional automation system used for control of an industrial plant consists of sensors connected to the plant, data acquisition devices, interface racks, actuators, cables and wires for transmission of analogue quantities, microprocessor-based controllers and a platform for operator intervention. The controllers, which are required to operate online in real time, are usually connected to plant equipment through relatively short-length cables/wires or optical fibres, designed with consideration of signal distortion, noise interference and cable reliability. Therefore, the controllers are distributively installed within a limited distance in the plant and if there are a large number of pieces of plant that undertake a variety of tasks within different time scales then the controllers are generally uncoordinated. For a complex industrial plant, such as a power system substation, chemical plant or steel manufacturer, it is difficult to connect various pieces of equipment, data acquisition devices, interface racks, actuators and controllers to central platforms and it is impossible to network these items horizontally and vertically within a hierarchical structure. The network would be very complex, as a huge number of cables and wires are used for a variety of purposes.

A modern power system contains a large number of monitoring and control devices, for example, the National Grid Transco (NGT) in the UK operates a transmission network of 244 substations at 275 kV or 400 kV, with a further 82 substations at 132 kV and below [1]. Whereas previously each substation had a centralised control system, modern substations are being equipped with many distributed Intelligent Electronic Devices (IEDs) performing various tasks [2]. Each of these IEDs is capable of sending and receiving data, often via a network, resulting in a large quantity of data becoming available. It is claimed that utilities are "among the largest users of data" and "the largest users of real-time data" [2]. However, engineers now have more data available than they are capable of managing in the time available to them [3]. In order to manage this amount of data, and allow utility engineers and management to make use of it in an appropriate manner, various systems and architectures have been proposed and developed which aim to integrate the data from

different IEDs and make it available to users [2]. Many of these (for example
[4, 5, 6]) are based on client−server methodologies and protocols such as
Hypertext Transfer Protocol (HTTP).

In comparison to client−server and object-oriented systems, multi-agent
systems have several claimed advantages. Jennings [7] states that "the natural
way to a complex system is in terms of multiple autonomous components that
can act and interact in flexible ways in order to achieve their set objectives",
and also that agents provide a "suitable abstraction" for modelling systems
consisting of many subsystems, components and their relationships. Ferber [8]
describes how agents, as a form of distributed artificial intelligence, are suitable for use in application domains which are themselves widely distributed.
The modern power grid, with substations distributed throughout a wide area,
falls into this category of systems.

This book describes a substation automation architecture based on the
multi-agent systems methodology. This chapter begins by presenting the historical background of substation automation systems, along with the newer
network-based approaches and architectures. A novel concept of e-Automation
is also introduced, and the main contributions of the book are presented.

1.1 Industrial Automation

The term *industrial automation* covers a range of systems used to improve the
productivity, safety or product quality of an industrial concern [9]. Normally,
the main function of any industrial automation system is to control a process
being performed.

Industrial automation systems may be applied to a wide variety of industries, which may be approximately grouped into the two categories of *process
industries* or *continuous process industries*, such as electric power systems
and other utilities, and *discrete manufacturing industries,* which include industries in which individual items, such as motor vehicles or electronic goods,
are produced. The type of automation system that is appropriate to a process
industry may differ from the type of automation system that is appropriate
to a discrete manufacturing industry [10].

There are several models of industrial automation systems in common
usage. One of the more well known is the *Computer Aided Manufacturing
(CIM)* [1] *pyramid* model, in which the system is viewed as a series of layers,
ranging from low-level data acquisition and control functions to high-level
functions such as plant and process management [9]. This model is shown in
Figure 1.1.

A more detailed model of industrial automation systems is the CIM reference model developed by the International Purdue Workshop on Industrial

---

[1] In the 1980s, the term *computer integrated manufacturing (CIM)* was used to
describe a range of industrial automation systems based on computers.

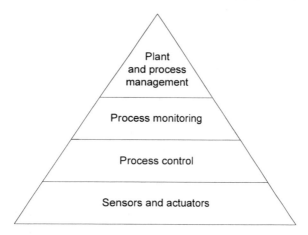

Fig. 1.1. "CIM pyramid" model of an automation system

Computer Systems [11]. This model describes in a generic fashion the tasks, and to some extent the implementation, of an "integrated information management and automation system", with most of the description aimed at the manufacturing industries. As with the CIM pyramid model, the outline structure of this model is hierarchical, consisting of a number of levels. The five levels included [11] are:

- Operational management: this level is responsible for overall production scheduling, coordination and reliability assurance.
- Section/Area: the duties of units at the area level include production scheduling, maintenance and local cost optimisation for a particular area. The area level is also responsible for generating production reports and analysis of operational data.
- Supervisory control: units at the supervisory control level are responsible for responding to emergency conditions in the plant, optimising the operation of the controllers and maintaining "data queues" for lower-level units under their control.
- Control level: the duties of this level include direct control of plant, human/machine interface and the collection of information for transmission to higher levels.
- Equipment: this level includes individual machines, sensors and actuators.

However, it is stated that "the number of levels used in a factory model is arbitrary", and that the six levels serve only to assist the standardisation process [11]. Within the hierarchical structure, control flows either within a level or downwards through the levels, and information/data flows upwards [11].

It has been suggested [11] that the usual method of implementation of a CIM system is to use a "hierarchy of separate computers". In support of this statement, the authors of [11] state that hierarchical systems provide the ability to implement distributed control of the plant, with each computer controlling a local area, and that the hierarchical structure follows the usual human management structure of a plant. It is conceivable that a multi-agent approach might be one way to implement such a hierarchical structure (see the description of hierarchical multi-agent systems, Section 2.6).

1.2 Automation Systems in Electricity Transmission

Electricity transmission networks consist of a number of substations interconnected by transmission lines. Each substation contains transformers, switchgear (disconnecters and circuit breakers) and other items of plant and protective equipment [12].

In the transmission industry three types of automation system are used: Supervisory Control And Data Acquisition (SCADA) systems, Energy Management Systems (EMS) and Substation Automation Systems (SAS) [13]. These form a hierarchical structure, with the EMS on the top level, the SCADA system directly subordinate to the EMS, and, at the lowest level, the individual SASs of each substation.

Energy Management Systems

An energy management system controls the overall operation of the power system [14]. The components of an EMS are [15]:

1. Network analysis, including state estimation, load flow optimisation, dispatching and voltage control.
2. Generation scheduling and control.
3. Data storage and retrieval.
4. SCADA, including data acquisition, alarms and Human–Machine Interface (HMI).

Therefore, a SCADA system forms one component of an energy management system. In the conceptual model of an EMS presented in [15], the SCADA system connects the plant, or "external equipment" to the database, with the other two functions (network analysis and generation scheduling) operating on the stored data.

SCADA Systems

The SCADA system of a power system, shown in Figure 1.2, is responsible for data acquisition, HMI functions and alarm/event processing [15]. In a SCADA system, a centrally located master computer is connected via some form of network to a number of Remote Terminal Units (RTUs), located in the substations and connected to the local substation automation system or Substation Control System (SCS). The master periodically polls the RTUs to retrieve status information, and can send commands back to the RTUs for execution by the substation automation system.

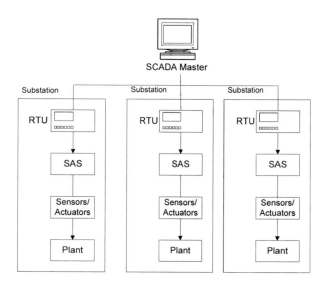

Fig. 1.2. Supervisory control and data acquisition (SCADA) system

Substation Automation Systems

A SAS is used to monitor and control a single substation, and to collect data for transmission to the overall SCADA system [16]. The main functions required of an SAS are *control* and *protection*. Control involves the operation of the plant, either locally or from a remote location. Protection, normally performed at the lowest level of the system by a number of relays, prevents damage to the system in the event of a fault. Various protection functions are provided, such as overcurrent, overvoltage and thermal overload. A more modern SAS will also provide a graphical human−machine interface, condition monitoring and historical data logging [5], along with remote access for information.

1.3 Network-based Power System Automation

Traditional substation automation systems have a number of drawbacks. The interoperability of devices is hampered by "an excess of incompatible hard-

ware interfaces and protocols" [17], and more access to substation information
is required in order to make business decisions [2]. In recent years, a number of
new architectures and products have been developed which aim to address one
or more of these drawbacks. In particular, in the last 10 years there have been
moves to integrate Local Area Networks (LANs) into industrial systems. This
has led to the development of several industry-oriented network architectures,
such as fieldbuses, Modbus^{TM} and Profibus$^{\circledR}$ [18]. Also, a number of systems now use standard Ethernet networks [19, 20] and the TCP/IP Internet
protocol suite.

The use of Ethernet and TCP/IP in an automation system has an important advantage in that it allows control systems to be connected directly
to office and enterprise networks, which themselves usually use Ethernet
and TCP/IP. This means that data collected by the control system can be
shared with other systems such as databases and Enterprise Resource Planning (ERP) systems, and can be viewed by users who are not located at the
production site. It is also possible to make process information accessible over
the World Wide Web [6], which removes the need for a specific client program to be installed on a user's system as data can be viewed with a Web
browser. However, even when a Web browser is used it may still be necessary
to download a large software component to the user's machine to allow online
monitoring data to be displayed.

A recent system installed in an Australian substation, described in [16],
consists of a number of distributed RTUs connected by a fibre-optic ring
network. The RTUs communicate with "intelligent relays" and input−output
devices, and there are duplicate HMIs and communication links to the SCADA
system at the control centre.

Client−server and Distributed Object Systems

Many current networked automation systems employ client−server technology, similar to that shown in Figure 1.3. A number of devices, either simple
sensors or actuators or, more commonly, IEDs, which incorporate an embedded processor, are connected via a network, such as Fieldbus^{TM} or Ethernet,
to one or more servers. These servers normally run the applications which
perform the centralised control functions of the system, including supervisory control, alarm and event management and data storage. A number of
clients can then connect to the servers either using client programs or a Web
browser[4, 19].

A client−server system employs a "request−reply" method of interaction.
The client system sends a request to the server, which must then carry out
a specified action, such as retrieving a Web page for display, and transmit a
reply to the server. An example of a client−server protocol is HTTP, which
is used to display information stored on a Web server.

Distributed object systems, such as CORBA$^{\circledR}$ [21, 22], Microsoft D/COM
and Java^{TM} RMI, provide an alternative methodology for distributed pro-

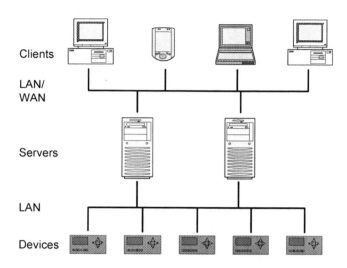

Fig. 1.3. Client−server automation system

gramming. A distributed object system extends the object-oriented programming methodology to cover objects located on multiple computers. Server objects may expose a number of methods via some standardised interface description, which other objects may invoke by sending a message to the server [23]. Unlike agent-based systems, the interface of a distributed object is pre-defined (in terms of its methods), and the interactions are normally synchronous (an object must wait for a method to complete before it can continue execution).

Relevant Standards

To address the problem of interoperability between IEDs produced by different manufacturers, a number of standards governing substation communications have been developed, or are under development. The IEC 61850 standard [24] specifies a model for intra-substation communications. The standard is split into several parts, of which IEC 61850-5 defines the basic structure for the system. The system is broken down into a number of functions, which are the various tasks that it must perform. Each function is performed by one or more Logical Nodes (LNs), which are situated within some physical device. However, the standard does not specify how the functions should be allocated to devices. An LN is defined "by its data and its methods". IEC 61850-5 also defines a number of interfaces between these logical nodes.

Data exchange in IEC 61850 is based on Pieces of Information for Communication, or PICOMs. A PICOM consists of an item of data to be transferred, along with information about its data type, permissible transmission time, source logical node and destination logical node or nodes. The standard also includes an object model (based on GOMSFE: Generic Object Model for Substation and Feeder Equipment) [25], and an event system for use in communications between protection devices (GOOSE: Generic Object Oriented Substation Events) [26].

Earlier substation communications standards include the Distributed Network Protocol (DNP) [27] and Utility Communications Architecture (UCA) [28].

1.4 e-Automation

A novel concept of e-Automation has been developed at the University of Liverpool. The concept defines a new generation of automation systems. These systems use the latest networking and agent technologies for information management, condition monitoring, and real-time control of a wide range of distributed industrial systems.

An e-Automation Laboratory has been established to undertake the research and development of e-Automation systems. This international flagship laboratory is also supported by NGT. The laboratory contains a Real-time Simulator, a range of hardware of microprocessors and embedded systems and data acquisition devices, real-time automation platforms, comprehensive software development systems and three IP networks, including a wireless local area network, which are used to undertake research in the area of network-based industrial automation.

The laboratory is established based on an IP network-based informatics grid, IP-based automation devices, local and global servers and platforms, with provision of hardware and software development tools as shown in Figure 1.4. The laboratory concentrates on capitalising the latest Internet, optics and microprocessor technologies for development of e-Automation systems, which can be used for information management, condition monitoring and control of various industrial systems. This development involves several technological and methodological aspects such as IP technology, agents technology, informatics grid, intelligent fibre-optic-based modules, microprocessor-based devices, Internet enabled distributed database, information management, data acquisition, synchronisation of data transportation, Global Positioning System (GPS) technology, network security, network computing, information processing and platform technology. Among these aspects, the IP Technology, the Agents Technology and the Technology of Advanced Modular Systems are regarded as the key to develop a basic frame of an e-Automation system, achieve the network communications and implement the operation and management of the e-Automation system.

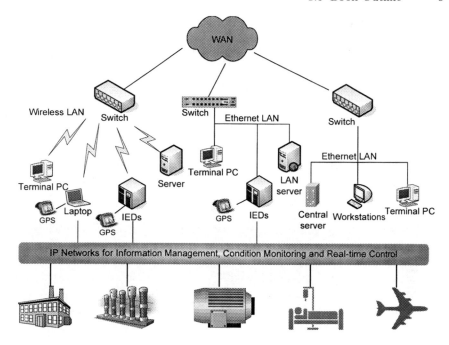

Fig. 1.4. e-Automation for large-scale distributed complex systems

1.5 Book Outline

The main contribution of this book is to examine the use of multi-agent, mobile agent and other computing technologies in the context of power system automation. The main aims are to determine the applicability of existing computer and information systems techniques to this domain, and to examine how different methodologies such as multi-agent systems and mobile agents can be combined into an integrated automation system. To this end, the following contributions have been made:

- A multi-agent architecture for power system information management, monitoring and control has been developed. By using agents to represent components of the automation system, it is possible for the architecture to more closely match the distributed nature of the system, and the flexibility of the system is increased by allowing components to be added and removed at runtime.
- As part of the development of the information management architecture, the representation of various aspects of power system knowledge in a form suitable for use in multi-agent system communications has been examined, and various examples of this have been provided.
- A prototype implementation has been produced which demonstrates the feasibility of the architecture. The prototype consists of agents which per-

form the data acquisition, information management and remote control functions of an industrial automation system, for a single substation. Mobile agents are used for data analysis and report generation. The prototype also provides an extensible platform for further research into intelligent applications in the power system.

- The use of mobile agents for a number of applications in power systems, as part of the overall architecture, is discussed and evaluated. Experimental results are provided which suggest that for the applications of data analysis/report generation and remote control, it is possible for mobile agents to provide significantly increased performance compared to static agents or client−server systems. This is particularly true when the user is connected to the substation by a network of high latency or, in the data analysis case, low bandwidth.
- The functionality, performance and flexibility of the architecture have been evaluated, providing insight into the utility of the multi-agent approach to the design of industrial automation systems, and the design decisions involved in developing a multi-agent-based automation system architecture. Particular attention has been paid to the architecture of the data acquisition system due to its performance requirements, and various possible multi-agent configurations for data acquisition have been evaluated.

This Chapter

has introduced current industrial and power system automation systems and different techniques used in the construction of such systems, as well as multi-agent systems and mobile agents.

Chapter 2: Agents, Multi-agent Systems and Mobile Code

gives a brief introduction to the fundamentals of the agent systems. Following the overview of the definition of intelligent agents, four distinct agent architectures are viewed. This chapter also introduces the FIPA standards specifications and mobile agent standards. The mobile agent technology and its applications, development of multi-agent systems, agent programming languages and design of middle agents are discussed. Finally, a set of agent architecture applications are presented.

Chapter 3: An Agent-based Architecture for Power System Automation

gives an overall view of the proposed architecture. First, the tasks that must be performed by an industrial automation system are considered. These tasks are then mapped into a multi-agent system, in which each task is performed by an agent or a number of agents, also taking into account the physical structure of the power system. The collaboration between agents to perform the various tasks is then discussed. The representation of knowledge for agent communication is described and a basic ontology for automation systems is defined. Finally, the implementation of the agent platform, upon which the multi-agent system is constructed, is considered.

Chapter 4: Static Components of Architecture

provides a more detailed examination of the static agents that make up the architecture. The basic agent architecture used by all agents is presented, and the sensors, effectors and knowledge of each agent are then considered individually. Implementation issues relating to particular agents, for example the plant agents which control and monitor items of substation plant, are discussed in detail.

Chapter 5: Applications of Mobile Agents

examines the use of mobile agents and mobile code. Previous research into mobile agent performance is discussed, and used to define a performance model for the mobile agent applications described. Experiments in remote control and data analysis are presented, including agent algorithms and detailed performance results. Other proposed applications such as remote monitoring are also considered.

Chapter 6: Multi-agent-based Substation Information Management System

describes a substation information management system implemented using the proposed architecture. This system is used to provide online monitoring, historical data querying and analysis and remote control for a substation simulator provided by the NGT. The particular agents and ontology used in the implementation of the system are described and detailed examples of the system's use are given.

Chapter 7: Evaluation and Analysis

provides an evaluation of the whole architecture, using the criteria of functionality, modifiability, performance and reliability. The architecture is evaluated with reference to other industrial automation systems, criteria for substation automation systems and other possible software architectures. Variations on the proposed architecture are also considered with respect to their expected performance in data acquisition.

References

provides 166 relevant articles, reports and websites.

2

Agents, Multi-agent Systems and Mobile Code

2.1 Overview of Agent Technology

While there is no fixed definition of an *agent* (or *software agent*), the concept is typically used to refer to software components that have their own thread of control (and hence may act autonomously), and are capable of sensing and reacting to changes in some environment. Often software agents have other properties, such as the ability to communicate with other agents. Recently, software agents have become widely used in the modelling of complex, distributed, problems [7]. This section discusses in more detail the various types of agents that are in use, along with multi-agent systems, which are systems in which agents interact in order to solve some problem or achieve a set of goals, and mobile agents, which are agents capable of moving from one server to another during their execution.

2.2 Intelligent/Autonomous Agents

2.2.1 Definitions

There are a number of definitions of an *intelligent agent*. One of the more widely used is that put forward by Wooldridge and Jennings [29], which defines an agent as a system that "enjoys the following properties:

- autonomy: agents operate without the direct intervention of humans or others, and have some kind of control over their actions and internal state;
- social ability: agents interact with other agents (and possibly humans) via some kind of agent-communication language;
- reactivity: agents perceive their environment (which may be the physical world, a user via a graphical user interface, a collection of other agents, the Internet, or perhaps all of these combined), and respond in a timely fashion to changes that occur in it;

- pro-activeness: agents do not simply act in response to their environment they are able to exhibit goal-directed behaviour by taking the initiative."

In the same paper, Wooldridge and Jennings described the notion of a *strong agent,* used by researchers in the artificial intelligence field, which is an agent that "is either conceptualised or implemented using concepts that are more usually applied to humans". An example of a strong agent is one based on a *mental state* described in terms of beliefs, desires, intentions and commitments. In contract to the weak notion, the characteristics of the *strong agent* are listed below:

- clearly identifiable problem-solving entities with well-defined boundaries and interfaces;
- situated in an environment, agents perceive through sensors and act through effectors;
- designed to fulfil a specific purpose, the agent has particular objectives to achieve;
- autonomous, the agent has the capability of control, the effectors depend on both its internal states and behaviours;
- capable of exhibiting the problem-solving behaviours in pursuit of its design objectives. The agent needs to be both reactive (able to respond in a timely fashion to changes that occur in its environment) and proactive (able to act in anticipation of future goals).

Strong agents can also be known as *cognitive* agents, in comparison to simpler *reactive agents,* which are agents that act only in response to changes in the environment [8]. One of the simplest type of reactive agent is an agent having a series of IF-THEN rules, mapping from input states to actions.

2.2.2 Intelligent Agents in Information Processing and Problem Solving

Over the past few years, a revolution has taken place in information generation and dissemination. Global distribution of information can now be made available via the Internet very easily. However, the problems of using the Internet and Web to retrieve information of interest are growing, and include:

- inexhaustible pool of information;
- distributed, heterogeneous nature of information and information services;
- difficulty of retrieving relevant information and more time spent searching for information.

Useful tools are needed for users to reactively and pro-actively search, retrieve, filter and present relevant information. Querying and integrating heterogeneous data from distributed sources has been a focus for research in recent years, as the notion of agent software assistants that help users to achieve productivity gradually becomes the trend in network computing. If given "intelligence", agents could be extremely useful, which means that agents can be trained to learn user actions and preferences, and perform appropriately to similar actions or specific preferences next time.

Intelligent software agents have obtained encouraging results in solving the problems of current (threat of) information overkill and information retrieval (IR) over the Internet. The complexity of information retrieval and management would be mitigated if intelligent means could be provided. Take "ACQUIRE" [30] for example, the following functions of information retrieval should perform in theory:

- goal-oriented: ACQUIRE is only looking for what the user wants, with consideration of how to most efficiently satisfy the request;
- integrated: ACQUIRE provides a single, expressive, and uniform domain model from which queries may be posed;
- balanced: ACQUIRE attempts to balance the cost of finding information on its own using planning and traditional database query optimisation techniques.

Furthermore, the following three stages are implemented by ACQUIRE

- respond to a query from a user and decomposes it accordingly into a set of sub-queries with site and domain models of the distributed data stores;
- generate an optimised plan intelligently for retrieving answers to these sub-queries over the Internet and deploys a set of intelligent mobile agents to delegate these tasks;
- merge the answers returned by the mobile agents and then return them to the user.

An example of agent coordination for problem solving is the Reusable Task Structure-based Intelligent Network Agents (RETSINA) [31]. The multi-agent-based RETSINA infrastructure has been developed at the Carnegie Mellon University in Pittsburgh, USA.

The RETSINA framework has been implemented in Java programming language. It is being developed for distributed collections of intelligent software agents in problem solving. The system consists of three types of reusable agents, which can be adapted to address a variety of domain-specific problems. Three major intelligent agents are integrated into the system architecture:

- an interface agent for interacting with users to receive the requirements and display the outcomes;
- a task agent, as an assistant to the users, which performs tasks according to the problem-solving plans, and queries and exchanges information with other agents;
- an information resource agent for intelligent access to a heterogeneous collection of information sources.

2.3 Agent Architectures

Four basic agent architecture categories and key examples are reviewed in this section; deliberative architectures, reactive architectures, learning-based architectures and layered architectures.

2.3.1 Deliberative Architectures

There are several different architectures for intelligent agent implementation. A well-known cognitive architecture is the belief-desire-intention (BDI) architecture [32, 33], in which the agent's knowledge base is described by a set of *beliefs* (those facts which an agent considers to be true), *desires* (those conditions which the agent wishes to bring about), and *intentions* (actions which the agent has committed to perform). These are explicitly represented in the knowledge base; for example, the Procedural Reasoning System (PRS) implementation [33] represents beliefs and goals as ground literals (sentences containing no implications, binary operators or variables) in first-order logic [34]. As described in [34], a BDI agent is capable of both reactive and deliberative behaviour. On each execution cycle of the interpreter, the agent retrieves new events from the environment. It then generates a set of *options*, which are plans or procedures that the agent is capable of carrying out, both in response to events and in order to achieve its goals. The agent will then execute, or partially execute, one or more of the selected options. This process is repeated for the agent's lifetime.

The BDI architecture was originally developed in the early 1990s, based on a model of human reasoning developed by Michael Bratman. One of its most famous early implementation was the PRS system, developed by Ingrand, Rao and Georgeff [33]. Rao and Georgeff, among others, have published a large number of papers concerning the BDI architecture and its implementation. The basic structure of a BDI agent is shown in Figure 2.1.

BDI Algorithms

Singh, Rao and Georgeff [34] give a set of algorithms for a basic BDI interpreter. The main interpreter loop is as follows:

```
BDI-Interpreter()
initialise-state();
do
   options := option-generator(event-queue, B, G, I);
   selected-options := deliberate(options, B, G, I);
   update-intentions(selected-options, I);
   execute(I);
   get-new-external-events();
   drop-successful-attitudes(B, G, I);
```

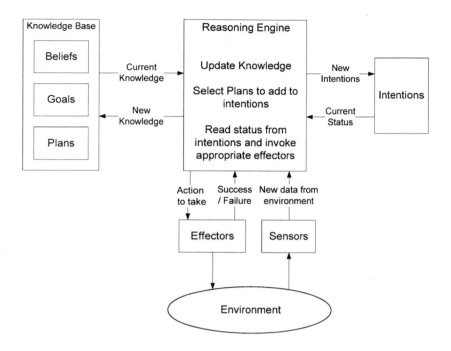

Fig. 2.1. Structure of a BDI agent

```
   drop-impossible-attitudes(B, G, I);
until quit;
```

In addition, the option generation procedure is defined as follows:

```
option-generator(trigger-events)
options := {};
for trigger-event ϵ trigger-events do
  for plan ϵ plan-library do
    if matches(invocation(plan),trigger-event) then
      if provable(precondition(plan),B) then
        options := options ⋃ {plan};
return(options)
```

The deliberation procedure can be implemented in a number of ways. In the PRS system, a procedure was implemented that allowed the user to include a set of *metalevel plans* that were invoked in order to select other plans to be carried out. The deliberation procedure was then as follows:

```
deliberate(options)
if length(options) ≤ 1 then return (options);
else metalevel-options := option-generator(b-add(option-set
```

```
(options)));
   selected-options := deliberate(metalevel-options);
      if null (selected-options) then
         return (random-choice(options));
else return (selected-options);
```

BDI Components

The mental state of a BDI agent has three main components: beliefs, desires (or goals) and intentions.

Beliefs

The beliefs of an agent are those things which the agent considers to be true. For example, I might believe that the circuit breaker "bolney/cb182" is open.

Singh, Rao and Georgeff [34] give a formal definition of BDI concepts based on modal logic. They state that:

- An agent can change its beliefs over time.
- If an agent believes a condition, then it believes that it believes that condition. If an agent does not believe a condition, it believes that it does not believe it.
- If an agent believes a condition, then it must not believe that it does not believe that condition.

In PRS and other implemented BDI systems [32], it is common to represent beliefs as *ground expressions* in predicate calculus, in order to reduce the execution time of the option generation and deliberation procedures. In this case, they consist only of a predicate symbol and a set of literals, and cannot contain variables or operators such as implication, conjunction and disjunction. Also, the beliefs represent only those facts that the agent *currently* believes. For example, the statement that the circuit breaker is open could be represented in the agent's knowledge base as:

```
status(bolney/cb182, open)
```

For reasoning about the behaviour of agents, Singh, Rao and Georgeff [34] use a modal operator Bel, and a relation B. This formalism is also used by Cohen and Levesque [35]. Suppose that x is an agent, and that p is a proposition. Then

```
Bel x p
```

means that x believes p. FIPA SL [36] includes the B relation, which has a similar meaning:

```
(B x p)
```

Desires and Goals

The desires of an agent are conditions that the agent would like to become true. Desires may be inconsistent with each other, and it may not be possible for the agent to achieve all of its desires. The agent's goals are a subset of its desires, which must be consistent and achievable.

In actual implementations it is possible to simplify the agent by having only goals and no desires.

Cohen and Levesque [35] define two types of goal: achievement goals and maintenance goals. An agent having an achievement goal in relation to a condition believes that that condition is not currently true, and desires to bring it about. An agent having a maintenance goal in relation to a condition believes that that condition is currently true, and desires that it should remain true.

Intentions

The intentions of an agent are those goals that the agent has committed to achieving, or actions that the agent has committed to carrying out. The commitment that an agent makes to its intentions is not indefinite. If an agent decides that one of its intentions is impossible, or that the justification for that intention has been removed, then it may drop or revise that intention.

Cohen and Levesque [35] state that if an agent intends to achieve a condition p, then:

1. the agent believes that p is possible;
2. the agent does not believe that he will not bring about p;
3. under certain circumstances, the agent believes that he will bring about p.

They also discuss whether or not an agent intends to bring about the expected side-effects of its intentions. They state that the usual view is that agents do not intend these side-effects, but that in their theory the side-effects are "chosen, but not intended".

2.3.2 Reactive Architectures

The subsumption architecture [37] is an example of a reactive architecture which does not employ an explicit knowledge representation. A subsumption agent consists of a number of concurrently-executing *behaviours* [38]. These are arranged in a number of layers, with lower layers representing simpler behaviours, which have a high priority, and higher layers representing more abstract behaviours, and having lower priority. Low-level behaviours are unaware of the presence of the high-level behaviours. It is therefore possible to construct an agent using the subsumption architecture starting with the lowest-level behaviour and working upwards, with the agent being functional, at least to a certain extent, after each layer is constructed. For example, Brooks [37] describes a mobile robot with a number of layers, performing tasks such as "avoid objects" (the lowest layer), "wander", *etc.*, up to "plan changes to the world" and "reason about behaviour of objects" (the highest layer).

2.3.3 Learning-based Architectures

Learning-based architectures, such as reinforcement learning [39], genetic programming [40], or inductive logic programming [41], may be used to enhance the performance of an agent. While it is possible to use learning to improve the capabilities of an agent using an architecture such as BDI (for example, [42] used machine learning methodologies to recognise plans being undertaken by other agents in a BDI architecture and [43] uses case-based reasoning in a BDI agent for information retrieval), it is also common to incorporate learning into a much simpler agent architecture. Learning agents have been applied in a number of domains, including user interfaces [44], telecommunications [45], control and robotics [46].

2.3.4 Layered Architectures

Layered architectures such as TouringMachines [47] and INTERRAP [48] are cognitive architectures consisting of one or more layers. According to Ferguson, the advantage of a layered architecture is that a layered agent, by having different levels of behaviour operating concurrently, is capable of reacting to changing circumstances while planning its future actions and reasoning about the behaviour of other agents. Both of the architectures mentioned have three layers: TouringMachines has a reactive layer, modelling layer and planning layer, while INTERRAP has a behaviour-based layer, local planning layer and cooperative planning layer. In TouringMachines, all three layers are connected to the agent's sensors and effectors. The three layers operate concurrently and are unaware of each other, while a control mechanism is used to filter the inputs and outputs and prevent conflicts. In INTERRAP, the sensors and effectors are connected only to the lowest layer (the behaviour-based layer). Activation requests are passed upward through the layers, and commitments are passed downwards. Unlike the subsumption architecture (which is a form of layered architecture), both TouringMachines and INTERRAP are based on explicit knowledge representation [47].

2.4 Standards for Agent Development

2.4.1 Foundation for Intelligent Physical Agents Standards

Foundation for Intelligent Physical Agents (FIPA) is an international organisation that is dedicated to promoting the industry of intelligent agents by openly developing specifications supporting inter-operability among agents and agent-based applications. FIPA standards was originally proposed in 1996 to form the specifications of software standards for heterogeneous and interacting agents and agent-based systems [49]. In the past a few years, FIPA has been widely recognised as the major standard in the area of agent-based computing. Many standard specifications have been developed, such as Agent Communication Language (ACL) and Interaction Protocols (IPs), *etc.* On 8 June 2005, FIPA was officially accepted by the IEEE Computer Society. Figure 2.2 shows the overview of the FIPA standards.

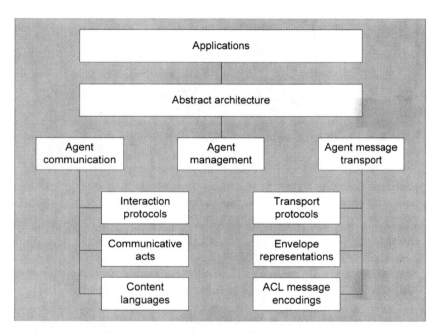

Fig. 2.2. Overview of the FIPA standards

FIPA Abstract Architecture

The FIPA Abstract Architecture specification (SC00001L)[1] acts as an overall description of the FIPA standards for developing multi-agent systems. The main focus of the FIPA Abstract Architecture is to develop semantic meaning message exchange between the different agents. It includes the management of multiple message transport and encoding schemes and locating agents and servers via directory services. Figure 2.3 demonstrates the FIPA Abstract Architecture mapped to different concrete realisations. In addition, it also

---

[1] All the specifications mentioned in this section are obtained from the website of FIPA organisation. http://www.fipa.org/repository/standardspecs.html

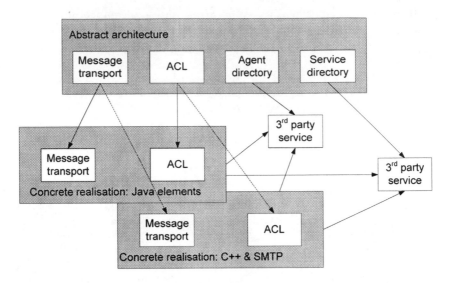

Fig. 2.3. FIPA abstract architecture mapped to different concrete realisations

supports mechanisms to create the multiple concrete realisations for interoperation. The scope of this architecture includes [50]:

- a model of services and discovery of services available to agents and other services;
- message transport interoperability;
- supporting various forms of ACL representations and content languages;
- supporting the representations of multiple directory services.

FIPA Agent Management System Standards

The FIPA Agent Management System specification (SC00023K) denotes an agent management reference model of the runtime environment that FIPA agents inhabit. The logical reference model is established for agent creation, registration, communication, location, migration and retirement [50]. The reference model includes a set of logical-based entities, such as:

- an agent runtime environment for defining the notion of agenthood used in FIPA and an agent lifecycle;
- an Agent Platform (AP) for deploying agents in a physical infrastructure;
- a Directory Facilitator (DF) which provides a yellow pages service for the agents registered on the platform;
- an Agent Management System (AMS) acting as a white pages service for supervisory control over access to the agent platform;
- a Message Transport Service (MTS) for communication between the agents registered on different platforms.

Figure 2.4 gives the FIPA agent management reference model constitution.

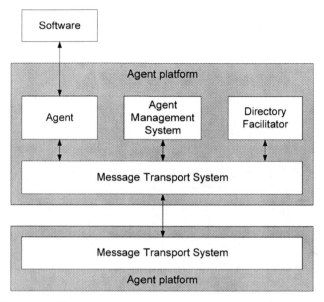

Fig. 2.4. FIPA agent management reference model constitution

FIPA Agent Message Transport Service

The FIPA Agent Message Transport Service specification (SC00067F), as part of the FIPA Agent Management specification, supports the message transportation between the interoperating agents. Two major specifications are involved, *i.e.* a reference model for an agent Message Transport Service (MTS) and the definitions for the expression of message transport information to an agent MTS [50].

A three-layered reference model is provided by MTS, *i.e.* the Message Transport Protocol (MTP) for physical messages transfer between two Agent Communication Channels (ACCs), the MTS which provides the FIPA ACL messages transportation between agents on the platform, and the ACL representations from both MTS and MTP. Figure 2.5 shows the FIPA message transport reference model. Additionally, other distinct components are involved in an agent MTS:

- two transport protocols, for transporting messages between agents using the Internet Inter-Orb Protocol (IIOP) and Hypertext Transfer Protocol (HTTP), specified by FIPA Message Transport Protocol for IIOP (SC00075G) and FIPA Message Transport Protocol for HTTP (SC00084F) respectively;

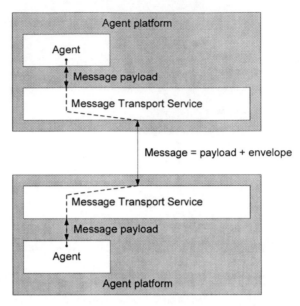

Fig. 2.5. FIPA message transport reference model

- two message transport envelope specifications, *i.e.* FIPA Agent Message Transport Envelope Representation in Extensible Markup Language (XML) Specification (SC00085J) and FIPA Agent Message Transport Envelope Representation in Bit-efficient Encoding Specification (SC00088D) which provide syntactic representations of a message envelope in XML form and bit-efficient form, respectively;
- three message representation specifications, *i.e.* FIPA ACL Message Representation in Bit-efficient Encoding Specification (SC00069G), String Specification (SC00070I) and XML Specification (SC00071E) for representing ACL syntax in a bit-efficient form, string form and XML form, respectively.

FIPA Agent Communication Standards

One of the most important areas that FIPA standardised is agent communication, which is the core category at the heart of the FIPA multi-agent system model. Four components are involved in FIPA Agent Communication specifications, *i.e.* Agent Communication Language (ACL) Message, Interaction Protocols (IPs) of message exchange, speech act-based Communicative Acts (CAs) and Content Language (CL) representations.

- FIPA ACL Message Structure specification (SC00061G) standardises the form of a FIPA-compliant ACL message structure to ensure interoperability.

- A number of different interaction message exchange protocols are dealt by FIPA Interaction Protocols (IPs) specifications, such as request and query interaction protocols, brokering and recruiting interaction protocols, subscribe and propose interaction protocols, *etc.*
- FIPA Communicative Act Library (CAL) specification (SC00037J) defines the structure of the CAL and the formal basis of FIPA ACL semantics.
- A set of languages used in FIPA Messages are denoted by FIPA Content Languages (CLs). For example,
  - a concrete syntax for the FIPA Semantic Language (FIPA SL) is defined by FIPA SL Content Language specification (SC00008I) for use in conjunction with the FIPA ACL;
  - FIPA Constraint Choice Language (CCL) Content Language specification (XC00009B) allows agent communication to involve exchanges about multiple interrelated choice;
  - FIPA Knowledge Interchange Format (KIF) Content Language specification (XC00010C) expresses the objects and propositions as terms and sentences, respectively;
  - FIPA Resource Description Framework (RDF) Content Language specification (XC00011B) constructs components of FIPA SL in the resource description framework representation.

2.4.2 Mobile Agent Standards

The mobile agent technology and its implementation are retarded due to the lack of world-wide, accepted standards which cover the most key areas of mobile agents, such as, code, data relocation, communication, interoperability, infrastructure and security. Currently, there are two main mobile agent standards available, OMG MASIF and FIPA Mobile Agent Standard.

OMG MASIF

The Mobile Agent System Interoperability Facility (MASIF) [51] is a standard for mobile agent systems, which was developed by Object Management Group (OMG) in 1998. OMG MASIF aims in migrating mobile agents between agent systems with same profile (*e.g.* language, agent system type, authentication type and serialisation methodologies) based on standardised CORBA Interface Definition Language (IDL) interfaces. The conventional client–server paradigm and the mobile technology can be integrated seamlessly based on the MASIF standard, which is built on the top level of CORBA.

MASIF is capable of interoperability between the agent systems developed in the same programming language which standardises the following concepts:

- Agent Management. As a system administrator, it provides agent systems management for different types by standard operations. Therefore, the agents can be remotely created, started, suspended, resumed, and terminated.

- Agent Transfer. Agent applications can freely move between the agent systems of different types. The agent class files are received and fetched between different platforms by agent transport methods, and available for agent management functions and execution subsequently.
- Agent and Agent System Names. The syntax and semantics of the agent and agent system names can be standardised by MATIS, which allows them to identify each other.
- Agent System Type and Location Syntax. If the agent system type is not suitable for the agent, the agent transfer will not happen. In order to locate each other, the location syntax is standardised by MATIS. In a mobile agent environment, agent tracking addresses the agents location, such as remotely query agent directories on different platforms.

However, the local agent operations such as agent interpretation, serialisation and execution are not be standardised by MASIF. In order to address interoperability concerns, the interfaces have been defined and addressed at the agent system rather than at the agent level. Two sets of interfaces constitute MASIF, such as, MAFAgentSystem and MAFFinder. So that, the specified remote operations are permitted to carry out transparently.

FIPA Mobile Agents Standard

FIPA standardised agent mobility in the FIPA Agent Management Support for Mobility specification (OC00005A) in 1998. However, this specification was not developed after 1999.

Two types of mobility are concerned in the FIPA 98 specification, mobility in devices and software. A series terms related to mobility are defined, such as the notions of mobile agents, migration, stationary agents, *etc.* In this specification, agents are allowed to take advantage of mobility by specifying the minimum requirements and technologies. Furthermore, combining with other FIPA specifications, a wrapping mechanism for existing mobile agent systems is proposed to promote interoperability.

Along with FIPA being officially accepted by IEEE Computer Society, a new working group was appointed in September 2005 to resume work on standards for mobile agents. The primary objective of this group is to improve and extend the existing specifications for agent mobility. The latest research results and experiences acquired from existing implementations and evaluation will be involved into the new specifications. Moreover, the number of reference implementations of protocols in software components will be developed as agent toolkits. The new specifications will be defined for communication interfaces and network protocols. According to the agenda, the Standard IEEE FIPA Mobile Agent specification will be reported in December 2006.

2.5 Mobile Agent Technology

Mobile agent systems are systems which involve the transfer of a currently executing program, known as a *mobile agent*, from one location to another. Fuggetta, Vigna and Picco [52] state that "in the mobile agent paradigm a whole computational component is moved to a remote site, along with its state, the code it needs, and some resources required to perform the task." Mobile agents were first discussed in the early 1990s, and applications to a wide range of areas have been proposed or implemented. For example, [53] describes a distributed calendar application implemented using mobile agents, [54] describes a military information retrieval application and [55] describes the application of mobile agents to network monitoring.

There are a number of reasons why mobile agents might be used in any particular application:

- Mobile agents can provide performance improvements by reducing network load [54, 56].
- Using mobile agents can allow servers to be made more flexible, with components being added and removed at runtime [57].
- Mobile agents permit disconnected operation, in which a client can "launch" a mobile agent into the network, disconnect, and then reconnect to retrieve the results of the mobile agent's task [58].

However, because of concerns regarding mobile agent security and a lack of motivation to deploy mobile agents, there has so far been limited use of mobile agents in real applications [59]. The mobile agent security issue consists of two problems: protecting a host and its data from a malicious agent or other attacker, and protecting an agent and its data from a malicious host or another agent [60]. While the first of these problems may be, at least partly, solved, there is still ongoing research into the second [61].

2.6 Multi-agent System

2.6.1 Architectures

Several architectural styles have been used in the development of multi-agent systems. Shehory [62] describes four such organisations:

- Hierarchical multi-agent systems, in which agents communicate according to a hierarchical structure, such as a tree. A system such as the Open Agent Architecture [63], which uses brokers, is a hierarchical system, as each agent communicates only with a broker or facilitator agent. Shehory gives the disadvantage of such a system as the reduction in autonomy of the individual agents, as lower levels of the hierarchy depend on and may be controlled by higher levels. However, hierarchical architectures can greatly reduce the amount of communications required, and also the complexity and reasoning capabilities needed in the individual agents.

- Flat multi-agent systems, in which any agent may contact any of the others. These provide the greatest agent autonomy, but result in more communications between agents. Also, agents in a flat structure must either know the locations of their communications partners, or be provided with agent location mechanisms such as yellow pages services. Many smaller multi-agent systems appear to be developed using a flat organisation.
- A subsumption multi-agent system is a system in which agents are themselves made up of other agents. In this system, the subsumed agents are completely controlled by the containing agents. This is similar to the subsumption architecture for an individual agent. According to Shehory, the fixed structure of a subsumption multi-agent system provides efficiency but restricts the flexibility of the system.
- A modular multi-agent system is comprised of a number of modules. Each module normally employs a flat structure, while intermodule communications is relatively limited. A modular multi-agent system might be useful in a situation such as power system automation, in which each substation could be categorised as a single module. Most communications within a power system are either within a substation or between a substation and the control centre, and so this might be an appropriate multi-agent system structure.

2.6.2 Multi-agent Programming

Logic-based Agent Programming Language

Jason and the Golden Fleece of Agent-oriented Programming

Jason [64] was designed as an interpreter language which is an extended version of AgentSpeak(L) [65]. As a pure logic-based agent programming language, *Jason* is based on the BDI architecture and BDI logics [32]. *Jason* allows agents to be distributed over the network through the use of SACI [66]. It implements the operational semantics of AgentSpeak(L) and the extension of that operational semantics to account for speech-act-based communication among AgentSpeak agents as given in [67] and [68], respectively.

Besides interpreting the original AgentSpeak(L) language, some of the other features available in *Jason* are given below [69]:

- speech-act based interagent communication (and belief annotations on information sources);
- annotations on plan labels, which can be used by elaborate selection functions;
- the possibility to run a multi-agent system distributed over a network (using SACI);

- fully customisable (in Java) selection functions, trust functions, and overall agent architecture (perception, belief-revision, interagent communication, and acting);
- straightforward extensibility by user-defined internal actions, which are programmed in Java;
- clear notion of multi-agent environments, which can be implemented in Java.

Cognitive Agents Goal Directed 3APL

Another multi-agent programming language worthy of mention is Goal Directed 3APL [70]. This language is designed to separate the mental attributes (data structures) and the reasoning process (programming instructions), which is motivated by cognitive agent architectures. One of the main features of 3APL is that it provides programming constructs to implement mental attitudes of individual agents directly, such as beliefs, goals, plans, actions, and practical reasoning rules. The basic building blocks of plans are composed of actions, which fall into different categories including mental actions, external actions and communication actions. An agent's belief base as well as the shared environment can be updated and modified in the implementation of selection and execution of actions and plans, which are allowed in the deliberation constructions.

An individual 3APL agent can be implemented by the 3APL programming language. Moreover, the shared environment, which the agent can be performed on, is designed by Java programming language as a Java class. Besides interacting with the environment, a 3APL agent can contact with each other agent via direct communication. Additionally, a 3APL multi-agent system is composed of a number of concurrently executed agents. The 3APL platform is built for sharing an external environment of a group of agents. Therefore, the platform allows the implementation and parallel execution of a set of agents and achieves the function of the 3APL agent programming language.

Using 3APL, the agents can be implemented of communicating with each other, observing the shared environment, reasoning about and updating the states and executing actions in the shared environment.

Java-based Agent Programming Language

JADE - A Java Agent Development Framework

JADE (Java Agent DEvelopment Framework) is a Java framework for the development of distributed multi-agent applications [71]. The JADE platform is open source software delivered by TILAB (Telecom Italia LABoratories).

JADE can be seen as an agent middleware which provides a number of available and convenient services, such as DF which provides a yellow page service to the agent platform. In addition, server graphical tools for debugging and testing the agent programs are also supported by JADE. One of the main characters of the JADE platform is that it strictly adheres to the IEEE computer science standard FIPA specifications to provide interoperability not only in the platform architectures design, but also in the communication infrastructures. Furthermore, JADE is very flexible and compatible with many devices with limited resources, such as the PDAs and mobile phones.

The goal of JADE is to simplify development while ensuring standard compliance through a comprehensive set of system services and agents [71]. To achieve such a goal, JADE offers the following features to the agent programmer:

- FIPA-compliant Agent Platform. Three agents are automatically activated with the JADE platform start-up, which includes the Agent Management System (AMS), the DF, and the Agent Communication Channel (ACC);
- distributed agent platform. The agent platform can be split on different hosts. Each host just executes only one Java application. Therefore, parallel tasks are processed by one agent and scheduled in a more efficient way by JADE;
- rapid transport of ACL messages inside the same agent platform;
- graphical user interface for organising different agent and agent platforms.

Jadex: A BDI Reasoning Engine

Jadex [72] is a software framework conducted by the Distributed Systems and Information Systems Group at the University of Hamburg. Following the BDI model, Jadex focuses on creation of goal-oriented agents. The framework supports the development of a rational agent layer on top of the middleware agent infrastructure, such as the FIPA compliant JADE platform. Therefore, the extended BDI-style reasoning and FIPA compliant communication are combined together in Jadex.

As a BDI reasoning engine, Jadex addresses the conventional limitations of BDI systems. Four basic concepts are integrated into Jadex, which are beliefs, goals, plans and capabilities. The reaction and deliberation mechanism in Jadex is the only global component of an agent [69]. Four goals are supported by Jadex, such as, a perform goal directly corresponding to the execution of actions, an achieve goal in the traditional sense, a query goal regarding the availability of agents required information, and a maintain goal for tracking a desired state.

After initialisation, based on internal events and messages from other agents, the agent is executed by the Jadex runtime engine by tracking its goals. In addition, many pre-defined functionalities and third party tools are integrated into Jadex, such as the ontology design tool Protégé [73].

JACK Intelligent Agents^{TM}

JACK Intelligent Agents^{TM} [74] was designed based on the BDI model and supported an open environment for developers to create new reasoning models. The JACK Intelligent Agents^{TM} framework was developed by a company called Agent Oriented Software, which brings the concept of intelligent agents into the mainstream of commercial software engineering and Java [75]. JACK Agent Language (JAL) is based on Java, which provides a number of lightweight components with high performance and strong data typing.

In addition, JAL extends Java in constructions in three main aspects to allow programmers to develop the components, such as: [75]

- a set of syntactical additions to its host language;
- a compiler for converting the syntactic additions into pure Java classes and statements which can be loaded with and be called by other Java code;
- a set of classes called the kernel provides the required runtime support to the generated code.

2.6.3 Middle Agents: Brokers and Facilitators

In a system consisting of only a few agents, it is possible to hard-code knowledge about other agents in the system, or to use a broadcast-based protocol, such as the Contract Net Protocol [76], in order for agents to locate others that are capable and willing to co-operate with them. However, as the number of agents increases, it becomes more and more difficult for agents to locate others which provide services that they require. This problem can be addressed by the use of middle agents, of which there are two main varieties: *matchmakers* (or *facilitators*) and *brokers*. A *matchmaking* system uses a set of agents that can be used by others to register their services. Other agents use these matchmakers to find agents providing a service that they require. Once they have done so, they communicate directly with that agent. A simple example of a matchmaking system is the FIPA Directory Facilitator [77].

In a *brokered* system, there are a number of agents, known as brokers. As with a matchmaker, agents can register their services with a broker. However, when a client wishes to use a broker, instead of searching for a suitable service-providing agent, they submit their request to the broker, which then handles the locating of a suitable agent, and sends the request to that agent. The client agent never communicates directly with the service-providing agent.

Decker, Sycara and Williamson [78] analyse the relative advantages and disadvantages of the brokering and matchmaking approaches. Their results suggest that the amount of time elapsed between a request being made and the same request being fulfilled is lower in a brokered system than in a matchmaking system, and that a brokered system can respond more quickly to the failure of service providers. However, they also state that in a brokered system, the broker is a single point of failure, and therefore the whole system can fail if a broker does. They conclude that a hybrid system containing both matchmakers and brokers may permit the advantages of both methodologies to be realised.

Kumar, Cohen and Levesque [79] use a set of brokers which work together as a team to prevent problems which arise as a result of broker failures. Each broker registers its service-providing agents with the team as a whole. If a broker fails, the other agents in the team will attempt to connect to the agents that were served by that broker. Once an agent has been reconnected in this manner, the broker that did so announces this fact to the other brokers, which will then not attempt to connect to that agent. The broker team is also responsible for maintaining a set number of brokers in the system.

2.7 Agent Application Architectures

There has been much previous work in the field of application architectures for multi-agent systems. For example, the RETSINA [80] architecture is a 3-tier architecture consisting of user agents, wrapper agents representing information sources and "middle agents" which transfer data between the two. Wrapper agents both "agentify" the data sources, allowing them to be queried using the agent communication language and convert data from the data models (ontologies) used by the individual data sources into a global ontology used for querying. Therefore, the wrappers assist in integrating data from heterogeneous data sources. RETSINA has been applied to several problems including financial portfolio management and visit scheduling. A similar information architecture, consisting of an ontology agent and database agents, is described in [81]. These architectures provide a basis for the development of a multi-agent information management architecture. However, they do not include the other functions used in power system automation, such as information management and control. Therefore, it is necessary to significantly extend these architectures to include this functionality.

The InfoMaster^{TM} [82] system used the Knowledge Query and Manipulation Language (KQML) [83] to provide a uniform interface to a number of heterogeneous data sources. The architecture of infomaster^{TM} consisted of data sources, each having their own wrapper, the "Infomaster^{TM} Facilitator" which performed data integration, and various user interfaces such as a Web interface. Infomaster^{TM} used rules to translate from individual database schemas into a global schema. Similar techniques are used by many information integration systems, including the multi-agent information management system described in this book.

A number of multi-agent systems have been employed to handle various aspects of industrial automation. The ARCHON^{TM} project [84] developed an agent architecture which was used in a number of applications. ARCHON^{TM} agents consisted of two layers: the *ARCHON^{TM} layer*, which was responsible for local control, decision making, agent communications and

agent modelling, and the *AL-IS interface*, which handled communications between the ARCHON^{TM} layer and the "intelligent system" being wrapped by
the agent .

There are many applications of agent technology in the manufacturing industries. These are similar in some ways to applications in the process industries and utilities, but often focus on machine control and task allocation. For
example, the PABADIS project [85] aims to develop a system for agent-based
manufacturing. The PABADIS system contains agents representing machines
and products [86]. The machine agents register descriptions of their capabilities with a lookup service, which can be used by the mobile product agents
to locate machines capable of carrying out the tasks involved in manufacturing a particular product. The PABADIS architecture demonstrates the use
of a multi-agent system in an industrial process. However, the architecture
of a manufacturing system differs from that of a utility. A utility system is a
continuous process, whereas a manufacturing system contains discrete parts
and outputs. Also, the area covered by a distributed utility system is much
wider than a single factory. In this work it is hoped to make use of the general
agent-oriented principles used by PABADIS (use of directories and representation of components of the plant as agents) but to design an architecture
more suited to the power systems and continuous process field.

Bussmann and Schild [87] used a multi-agent system in the control of
a flexible manufacturing system. The system was used to manage the flow
of material between different machines, and to allocate tasks to machines. It
was applied to automobile manufacturing. The approach taken by this system
was based on auctions, in which workpieces auctioned off tasks to machines.
It was found that this system provided both improved throughput and increased robustness compared to traditional methods [88]. As with PABADIS,
the relevance of Bussmann and Schild's work is restricted by the fact that
their application is in the manufacturing domain.

Leito and Restivo [89] describe a multi-agent architecture under development for "agile and cooperative" manufacturing systems. As well as controlling
the manufacturing system, the architecture supports re-engineering of products. The agent architecture consists of Operational, Supervisor, Product,
Task and Interface agents. This architecture is also intended for use in manufacturing industries. However, it might be possible to use agents corresponding
to the operational agent (which Leito and Restivo define as corresponding to
the "physical resources") and supervisory agent in a utility system. Also, as
the paper states, "only preliminary results are presented", and further work
is therefore required.

Mangina *et al.* have also developed the condition monitoring multi-agent
system (COMMAS) architecture [90], which uses three layers of agent. At-
tribute Reasoning Agents (ARAs) monitor and interpret sensor data, Cross
Sensor Corroboration Agents (CSCAs) combine data from different sensors,
and Meta Knowledge Reasoning Agents (MKRAs) provide diagnostics based
on the information provided by the other agents. Mangina's architecture is

relevant to the power systems domain. However, it provides only the single task of condition monitoring, and does not contain information management and remote control or operation functionality.

These applications have been relatively successful, suggesting that the multi-agent approach is a promising method for the implementation of industrial automation systems. However, the previous work described does not provide a single architecture for providing all the functions required by a power system automation system, either because it focuses on a single application or because it is intended for use in manufacturing industries. The work described in this book is intended to provide such an architecture.

3

An Agent-based Architecture for Power System Automation

The increasing use of IEDs and networks in power system substations has led to the availability of a large amount of data and information of various types, and standard protocols such as IEC 61850 [91] have improved the interoperability of different devices. However, it remains difficult to effectively manage the amount of data produced [92], and to convert this data into knowledge to enable engineers to make use of it [3]. A framework is required to provide open access to substation information via the power company's wide area network and to integrate previously separate functions such as protection, control and information management [93]. Another drawback with current automation systems is that they are inflexible and cannot easily accommodate new requirements or changes to the substation plant and monitoring equipment. It is hoped that a new architecture might be able to address this shortcoming.

The client−server model, used by most current systems, is widely supported and therefore provides a simple means to develop a distributed application. However, it is more suited to centralised applications, in which one server serves a number of clients, or one client controls a number of servers, than true distributed applications [94], and is lacking in flexibility. Distributed object systems, such as CORBA®, provide a more suitable representation, in which the system is composed of separate objects. However, in a distributed object system objects do not usually have their own thread of control, which means that it is not possible for different parts of the system to act simultaneously. Also, message passing in a distributed object system is usually synchronous, which means that an object that invokes a method on another object must wait for that object to respond before it can continue with any other tasks that it is involved in. The multi-agent approach provides increased autonomy by giving each agent its own thread of control, and provides asynchronous message-passing. Multi-agent systems also provide a high-level communications language (Foundation for Intelligent Physical Agents Agent Communication Language or FIPA ACL [95, 96]) with a clearly defined semantics, which is useful in information management and integration. Therefore, the architecture described here applies the multi-agent systems approach.

This chapter describes in detail a proposed multi-agent software architecture for power system automation [97, 98, 99]. The overall structure of the system consists of WAN (wide area network) and LAN (local area network) components, as shown in Figure 3.1. The LAN component represents those components of the architecture that would be installed at a substation, while the WAN component represents those components of the architecture found at other locations, such as an office or on a client's computer. There are multiple LAN components, one for each substation, and there may also be multiple WAN components.

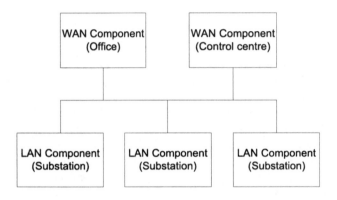

Fig. 3.1. Outline view of architecture

3.1 Agents in Power Systems

There have been several applications of agents and multi-agent systems to various areas of power systems.

3.1.1 Control

The ARCHON^{TM} system [100] described earlier was used to perform fault identification and service restoration in a power transmission network, and was installed in an actual control centre. Seven agents were used, based on both existing and new expert systems in the control centre. Each agent was responsible for a particular task, for example, blackout area identification or control system interface. The application of ARCHON^{TM} described was used only in the control centre, and was not a full substation automation system. However, the general principles of the ARCHON^{TM} system, including the use of wrappers, may be applied to the design of such a system.

Nagata *et al.* [101] applied a multi-agent system to the control of switching operations in a power system. The system consists of facilitator agents, equipment agents and switch box agents. Equipment agents represent transformers, buses and transmission lines and hold the operational status, maintenance time and continuation time of an item of equipment. Switch box agents represent circuit breakers or groups of circuit breakers known as "logical switches". Agent negotiation between facilitator agents and switch box agents is used to isolate parts of the system and enable maintenance work to be carried out, and for system restoration afterwards. The system was implemented in Java^{TM} and tested in simulation.

3.1.2 Negotiation and Pricing

Another application area of agents in power systems is to electricity pricing.

The system described in [102] uses negotiation between two agents, representing an independent system operator and a transmission company, to determine whether a transmission circuit may be operated in excess of its rated load and, if so, how much compensation should be paid by the system operator to the transmission company. The agents exchange proposals until a mutually acceptable solution is reached. The system was implemented in Java^{TM} and tested on a simulator. It was also used for decision making in other aspects of a power system [103].

A market-based system described in [104] makes use of a multi-agent system to optimise energy usage. The system contains device agents, representing consumers, service agents and utility agents, representing utilities. These agents participate in an auction mechanism to buy and sell power. It was found that the system was able to adapt to a change in price by utilities within a few auction steps. The claimed advantages of the system are that it is capable of dealing with large-scale systems, is flexible and adaptable, can be customised to individual requirements and may be used in a wide range of applications.

3.1.3 Information Management

Lucas *et al.* [92] describe a multi-agent system designed to provide document management, querying and decision support for the Iranian power industry. The system consists of interface agents representing users, resource manager agents representing information resources, and coordinator agents, which take requests from interface agents and pass them to the appropriate resource manager agents. These agents are used to connect several databases and sets of documents, including the Internet.

3.1.4 Condition Monitoring

There have been relatively few applications of multi-agent systems to condition monitoring in substations. However, Mangina *et al* have developed a multi-agent system, condition monitoring multi-agent system (COMMAS) [90], which has been used for condition monitoring of power plants [105]. By reasoning about what causes a gas turbine to move from one state to another, the agents are able to identify the causes of faults.

3.2 Tasks Performed

As discussed in the Introduction, a power system automation system must perform a variety of tasks, which operate at different timescales, have different characteristics and involve the transfer of different types of data. Figure 3.2 gives a set of common tasks which must be performed by an automation system, including the flow of data between the different processes of the system. This is based on the models of industrial automation discussed in Chapter 1.1, on the requirements for the SICAP substation automation system laid out in [93], and on experience with existing data acquisition and SCADA systems. The "system boundary" divides those processes which are implemented by the system from the external interactors (databases, users and plant). The processes shown are defined as follows:

User Interaction

involves taking commands and queries from the user, and translating them into an appropriate form for submission to the other components of the system. As shown in Figure 3.3, the task of interacting with the user involves handling queries, requests and online data display. The task of online data display is driven by events from the control system, while queries and requests originate with the user.

Intervention

provides the ability for a user to send commands which change the state of the system, for example, to open or close a circuit breaker or set the tap position of a transformer.

Output Data Interpretation

translates from a desired state, expressed in a command, to the appropriate setting (or sequence of settings) of digital and analogue outputs required to achieve this state

Data Acquisition

takes raw data from sensors and translates it into numerical values. This task corresponds to that of the "sensor process" described by Somerville [23].

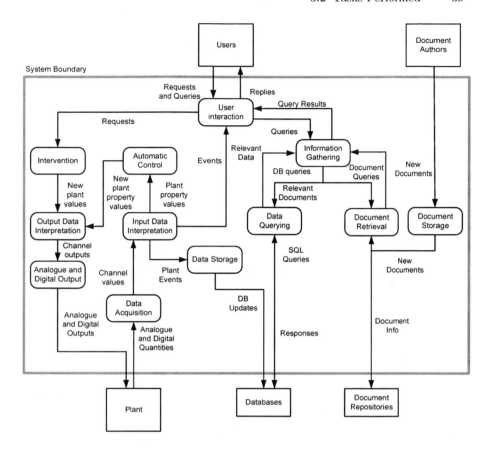

Fig. 3.2. Functional decomposition of power system automation system

Input Data Interpretation

takes the data gathered by the data acquisition process and converts it into a
machine-understandable representation of the state of the power system. This
task corresponds roughly to that of the "process data" process described in
[23].

Automatic Control

involves closed-loop or open-loop control of the system without the participa-
tion of a user.

Data Storage

takes event and other data from the plant (via the input data interpretation
process) and stores it as historical data in a database.

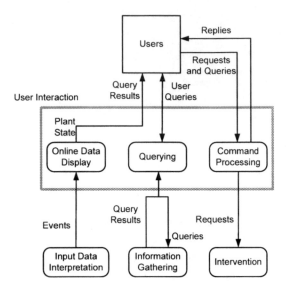

Fig. 3.3. User interaction

Data Querying

is used to retrieve stored data from a database. It takes queries from the information gathering module and transmits them to the database as Structured Query Language (SQL) queries, returning the results obtained.

Document Retrieval

provides an information retrieval facility for documents related to the power system. These include maintenance records for substation plant, technical documentation *etc.*

Document Storage

is used to allow users (either document authors or system administrators) to add new documents to a document repository.

Information Gathering

is responsible for taking queries from the user interaction component, or from other components, and retrieving data or documents that are relevant to those queries.

These tasks are now considered in three groups: data acquisition/control, information management and user interaction. For the purpose of this discussion, the output data interpretation, input data interpretation, automatic control and intervention tasks are performed by the data acquisition/control system and information gathering, data storage and document management by the information management system. As shown in Figure 3.4, all tasks, including information management, user interaction and data acquisition/control must be performed in a substation. However, wide area network locations (control centres/offices) perform only information management and user interaction tasks.

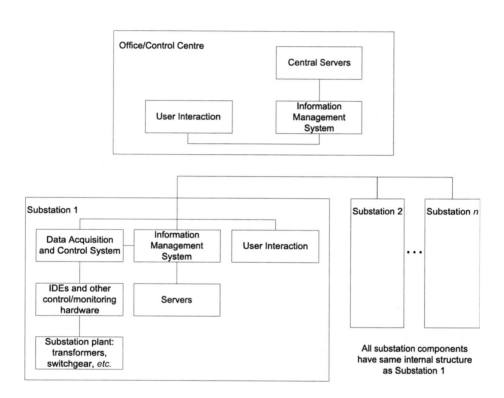

Fig. 3.4. Allocation of systems to substations and wide area network

3.3 A Multi-agent System for Power System Automation

This section describes one possible multi-agent system to fulfil the requirements of the tasks described in the previous section. In designing multi-agent systems, a common approach preferred by several authors in the field [106, 107] is to use a *physical decomposition* process, in which each object in the system is represented by an agent. This provides the derived system with a structure which is representative of the physical system being modelled and hence is easy to understand. Also, a physical decomposition may increase the ability of the system to cope with change. For example, if an item of plant is removed from a physically decomposed system it may only be necessary to alter the agent associated with this item of plant. In a functionally decomposed system, if an item of plant is removed, all the functional agents which make use of this item of plant will need to be altered. In the design of the multi-agent system described here, this principle is applied to derive plant agents which represent substation plant, device agents representing monitoring devices and user agents representing users.

However, physical decomposition may not always be the best approach to take. For example, in other work the functional approach is still used in agents such as brokers and mediators, even when the main multi-agent system uses a physical decomposition [107]. In the system described in this book, the main application of the functional decomposition approach is to mobile agents. There are separate mobile agents for monitoring, remote control and information gathering. This functional decomposition is useful for mobile agents because it allows relatively small mobile agents to be developed (the agents only have to perform a single specific task), which reduces the amount of data that must be transmitted across the network when a mobile agent is moved.

Another aspect to consider is the level of intelligence and autonomy to be exhibited by the agents. Again, a high level of intelligence might be detrimental to the mobile agents, as the large amount of code required would increase the transmission size of the agent. However, intelligence and autonomy might be beneficial to other agents in the system.

This section discusses the design of the system by considering its two main components: data acquisition/control and information management. However, these components are not completely separate. Some agents perform more than one role and are present in multiple components. For example, the plant agents both control the plant and act as a source of plant information to the information management system.

3.3.1 Agent Platform

An agent platform provides a basis for the implementation of a multi-agent system, and the means to manage agent execution and message passing. It is intended that the architecture should be implemented using an agent platform based on the FIPA specifications [77]. These specifications, being developed by an industry consortium of around 25 organisations, are the most prominent attempt to standardise multi-agent systems technology and are implemented in a number of publically available agent platforms. The specifications define an abstract agent platform, a number of services that must or may be provided by such a platform and a standard communications language. It is assumed that agents have access to the FIPA-specified directory facilitator (DF) service for publishing their capabilities and locating other agents capable of providing a specific service. The DF uses "service descriptions", defined in [77], to allow agents to identify the services provided by each other. In the system described here, an agent might need to use the directory to find out, for example:

- whether an agent is capable of providing the answer to a particular query. For example, suppose an agent knows that it can obtain the low-voltage current of a transformer SGT1 by querying channel 0 of device 0 on node IED1. The agent may then use the directory facilitator to obtain the identity of any agent that may provide it with the ability to query this channel;
- whether an agent is capable of performing a particular action.

Use of the directory facilitator permits agents to be added and removed at runtime, as an agent providing a service may be substituted with another agent providing the same service. A directory entry for an agent includes (among other items) the agent name, service name, service type, ontologies (data models) used, protocols used and a set of properties, which may be defined by the user, describing that service. Similarly to the mechanism used in [108] to register agents with brokers, in this system agents register the actual queries that they may answer with the directory facilitator. The service type "query-service" is used to denote a registration containing such information. For example, in the implemented device agent described in Chapter 6, the agent registers with the DF the actual information that it can provide using the system ontology, for example, in the FIPA Semantic Language (SL) [36] language, the following might state that the device agent can provide the value of a channel "tp1":

```
(any ?a (value tp1 ?a))
```

Another agent may then use a unification procedure (*e.g.* that found in Prolog) to match the information provided by information sources with a specified query[1]. This is only a basic mechanism, and in a complex system additional information (*e.g.* preconditions for a query to be answered) should be provided to enable agents to choose between two or more agents providing the same or similar services. In the system described here, a similar method is used for agents to register actions that they are capable of carrying out. The service type used for this is "request-service". All agents providing a service will register either a query service or request service. Agents may additionally register more specific services, for example, the ontology agent registers a "fipa-oa" service, which is a service defined in the FIPA standards for ontology agents [109].

---

[1] In most cases, it should be possible to ignore the *any, all* or *iota* part of the expression and match only on the inner expression. However, in order for the expression above to be a legal term in FIPA SL, there must be no free variables, and so *any, all* or *iota* must be included.

Inter-agent Communications

The use of the FIPA platform also provides a standard agent communication language, FIPA ACL [95]. FIPA ACL is a high-level agent communication language based on speech acts. An ACL message consists of an outer message structure, providing information such as the sender and receiver of the message, and a message content, expressed in some language understandable to both the sending and receiving agents [110]. Each message has a *performative,* or *communicative act,* which determines its type and hence the effect that it is intended to have on the receiver. The FIPA ACL specifications define a message structure [95], standard communicative acts [111] and interaction protocols. Wherever possible, the standard interaction protocols are used in this architecture. In addition, the language used for all inter-agent communications is the FIPA SL language [36].

3.3.2 Data Acquisition and Control System

An object model showing the components of a typical data acquisition system is given in Figure 3.5. The system consists of a number of data acquisition *nodes,* for example, PCs equipped with input or output hardware or stand-alone IEDs. If connected to a network, which we assume is the case for the development of this automation system, the node will have one or more network interfaces, represented by the *interface* class. This has various subclasses representing the different types of network available, including the *IPBased-Interface* shown on the diagram that represents an address on an Internet Protocol network, having a protocol name, host name or IP address and port number [112].

Each node is equipped with a number of *devices,* which represent a physical or virtual unit of I/O capability, for example, a PCI data acquisition card or a (non-modular) IED. It is possible that a node contains only one device—for example, a computer may have only one data acquisition card. However, nodes with 0 devices are not of interest to the data acquisition system, as no input or output facilities are provided. A device may contain a number of *channels,* each of which is capable of the input (*InputChannel*), output (*OutputChannel*) or both (*InputOutputChannel*) of a single analogue or digital value. Each channel has a data type, which defines the type of data input or output, for example, *Boolean* or *float.*

Each input channel measures a particular *property* of an item of plant via sensors and actuators (not shown on the diagram as they are external to the data acquisition system). For example, an analogue input device might use a thermocouple sensor to measure the oil input temperature of a transformer. It might be possible that a property is measured by more than one channel. However, because a channel is capable only of the input or output of a single quantity it is not possible for it to measure more than one property. If a property is directly controllable (*e.g.* the status of a circuit breaker), then it

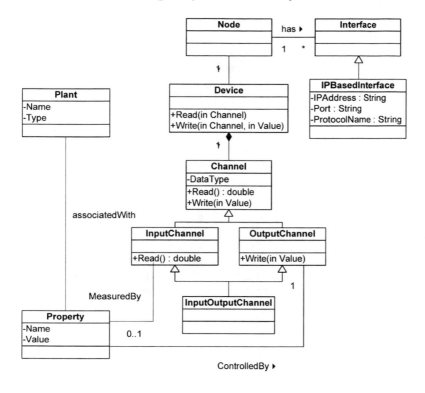

Fig. 3.5. Generic object model for data acquisition system (using UML class diagram notation)

is *controlled by* one or more output channels. The simplest method of control is that in which writing a value to a channel directly sets the value of the plant property, for example, writing a 0 to a digital output to close a relay. However, it is also possible that properties may be controllable, but are set indirectly via actuators. For example, writing a value to the "tap up" channel of a transformer would cause the tap position of that transformer to be increased by 1, but it is not possible to directly write a tap position value to the "tap position" channel.

Agents

From the model shown in Figure 3.5, and using the principle from [113] that agents should correspond to things within the problem domain rather than "abstract functions", it is possible to derive several possible agent communities, depending on the desired granularity of the decomposition. For example, considering the data acquisition system, which consists of nodes, channels and devices, should each channel be represented by a separate agent, or should a single agent represent all channels on a device, or even all channels on a node? Van Dyke Parunak [113] states that an agent should be "small in mass" (representing a small portion of the entire system), "small in scope" (having limited sensors and action capabilities) and "small in time" (able to remove or "forget" outdated information in its memory)—in other words, that in general agents should be kept small. This would suggest the use of separate agents to represent nodes, channels and devices. However, this may not always be simple to implement in practice, and would result in a very large number of agents, which might prove difficult to manage. Therefore, the proposed architecture specifies the use of one agent per device. However, because agents use DF entries to locate other agents capable of performing an action or responding to a query, it should be possible to substitute a channel agent for part of the functionality of a device agent without disturbing the rest of the system. All DF entries relating to the relevant channel would be removed from the description of the device agent, and put into the description of the channel agent. Therefore, "client" agents looking for, for example, the value of the channel would determine from the DF that the agent now providing that service was the channel agent rather than the device agent. An alternative architecture would be to have a hierarchical structure consisting of one device agent and multiple channel agents, with the device agent providing the external interface to the system. Other agents would not see the channel agents, and would pass requests to the device agent, which could then forward them to the appropriate channel agent.

Considering the plant and properties of plant, it follows from the principle of agents representing entities that each item of plant should be represented by a single agent. A property of an item of plant is not actually an entity, although it is shown on the object model as such to allow the link to be made between channels and properties. Each plant agent will be aware of all of the properties of its respective item of plant and, where appropriate, able to control them.

Figure 3.6 shows the derived multi-agent data acquisition and control system. The device agents are responsible for performing data acquisition and output on a single device. Plant agents obtain data from the device agents, and are responsible for converting that data into a representation of the current state of the relevant item of plant. They are also responsible for automatic control tasks relating to that item of plant, and for providing information to the information management system. Cooperative, distributed control schemes may be implemented by communications between the plant agents. For a typical substation, there will be plant agents representing transformers, circuit breakers, disconnecters and any other items of plant.

For most control and information management purposes, only the plant agents need be visible to the rest of the system, as users are normally interested in the functioning of the substation, rather than the details of the data acquisition system. However, as well as allowing the agents to be as "small" as possible, the use of intermediate data acquisition system agents between the

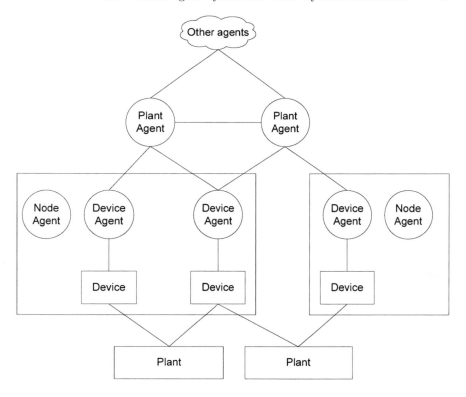

Fig. 3.6. Multi-agent system for data acquisition and control

plant and the plant agents, rather than solely using the plant agents, is necessary because there may not be a 1:1 mapping between plant and devices—a device may control or monitor multiple items of plant, and an item of plant may be monitored or controlled by a large number of devices. The device agents hide the different implementation details of these multiple data acquisition devices from the plant agents, making it possible to re-use the same plant agent implementation (with configuration changes) for different items of plant of the same type, even where the systems monitoring these items of plant are different. This architecture also simplifies the task of changing a data acquisition device or adding a new one. However, it is also conceivable that similar information hiding could be accomplished by the use of a library of drivers, each for a specific device but sharing a common interface, in place of device agents. Another advantage of the multi-layered structure is that functions such as the configuration of the data acquisition system (carried out by device agents) are separated from the control functions (carried out by plant agents). This should make the implementation of the individual agents simpler, as they have fewer tasks to carry out.

Based on the principle of "small agents" discussed in [113], it is possible to conceive that the agents representing complex plant items such as transformers, which have many components, should themselves be split into several "sub-agents". For example, a transformer agent might be responsible only for monitoring the windings of the transformer (voltage, current *etc.*) and would be associated with a cooling system agent, monitoring the cooling fans and oil temperatures, and a tap changer agent, responsible for the operation of the tap changer. It is difficult to evaluate whether or not this would be useful in practice, and for reasons of time it has not been implemented in the current prototype of the architecture. However, it is an issue for further research.

3.3.3 Information Management System and User Interface

The architecture of the multi-agent information management and user interaction system extends previous agent-oriented information system architectures such as RETSINA [80], which is described in Section 2.7. This three-tier architecture consists of user agents, wrapper agents representing information sources and "middle agents" which transfer data between the two. Wrapper agents allow heterogeneous data sources to be queried using the agent communication language (which in this system is FIPA ACL), and convert data from the data models (ontologies) used by the individual data sources into a global ontology used for querying [114, 115].

The architecture for substation information management, shown in Figure 3.7, contains three basic types of database: data logging databases, which are used for storing status and event information, "static" databases which hold configuration data, and the ontology database, which holds the system ontology (data model). Each database has its own database agent, with the ontology database having the ontology agent. Additional data is provided by plant agents themselves and by substation document repositories. The plant agents do not require wrappers, as they are already agents. The other sources of information are represented in the system by wrapper agents: the databases by database agents and the document repositories by document management agents.

A mobile server is a server which is added to the system for a period of time (perhaps to carry out data acquisition for an experiment) and is then removed. Data stored on a mobile server may be accessed via the multi-agent system in the same way as data stored on any other server, providing that appropriate agents are available on the mobile server. The type of agent to be used would be determined by whether the mobile server contained a database (in which case a database agent would be used) or a data acquisition system (in which case device agents and a node agent would be provided). In the second case, the mobile server should be integrated into the data acquisition/control system, and pass data to a plant agent, rather than into the information management system.

The information transport (middle) layer of the system is composed of brokers, task-oriented agents and mobile agents. Task-oriented agents use the services and information provided by the service agents (database agents, document agents and plant agents) to provide a specific service to user agents. For example, an alarm and event agent might use monitoring data provided by the device agents to generate alarms and events for viewing by users. These agents provide a convenient encapsulation of particular forms of data. It is also possible to conceive of a variety of different task-oriented agents to perform various functions such as modelling, prediction or decision support. These agents are task-oriented rather than physically oriented. The use of task-oriented agents permits a variety of services to be implemented by introducing new agents rather than by modifying a large portion of the existing system.

The user interface layer consists of user agents, with each agent representing a particular user of the system, and a HMI. The user agents perform data transformation between the multi-agent system and the HMI, similar to the function played by a wrapper agent for a data source.

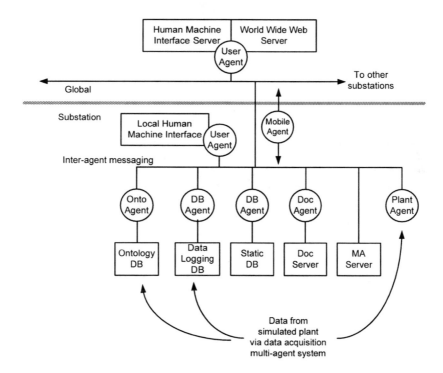

Fig. 3.7. Information management multi-agent system

Database Agents: Input and Output Agents or Input/Output Agent?

The database agent as described above must perform two tasks: data storage and data querying. However, it would also be possible to use separate storage and querying agents. The advantages and disadvantages of each configuration are now considered.

One Agent

The use of one database agent provides the ability to view the database, which is a single software system, as a single agent, and is therefore conceptually attractive. Also, the mapping rules (discussed in Section 4.1) used by a database agent to translate data from its database schema into the global ontology would only have to be stored within a single database agent. However, a major problem with the single-agent approach would be that the load on the agent would be increased as a result of its having to perform (possibly concurrent) data update and querying operations. It might be possible to mitigate this problem by the use of threading (with each agent having separate threads to handle update and querying). However, this increases the complexity of the agent implementation.

Two Agents

The use of two separate agents would allow the functions of data update and querying to be split between these agents, reducing the complexity of the individual agents, and making it simpler to modify either of the agents without affecting the other. However, a mechanism to share mapping rules between agents would have to be used, or else these rules would have to be duplicated in both agents. The use of two separate agents should also reduce the load on the individual agents, and it would also be possible to place the agents on separate computers, further improving performance and scalability.

3.3.4 Combined Multi-agent Architecture

By combining the components of the multi-agent system obtained in the preceding sections, an integrated architecture, the outline of which is shown in Figure 3.8 is obtained[2]. In the information management system, all agents may communicate with each other. However, the task agents may be situated in the substations, and in this case would not need to use mobile agents to retrieve data.

The interface between the data acquisition/control system and the information management system occurs at two points: the plant agents (for online

---

[2] In order to make the diagram readable, the architecture has been simplified in comparison to the diagrams of its individual components.

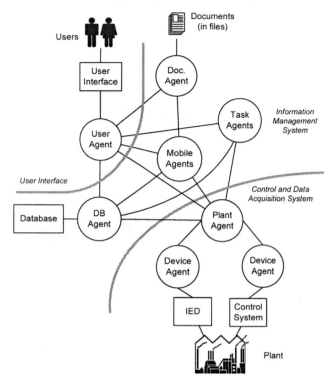

Fig. 3.8. Combined multi-agent system

monitoring and operator intervention) and the data logging database (for data querying). Mobile servers are not shown on the combined architecture, as it is envisaged that they would contain either a database and database agent, or a data acquisition system and plant/device agents.

Agent Collaboration

Figure 3.9 shows how the different types of agent in the system collaborate. The multi-agent system implements the generic system shown in the data flow diagram (Figure 3.2) and therefore the external interactions of the system (those which cross the system boundary) are the same.

Multiple Substations (Modular Architecture)

The descriptions of the architecture in this chapter have so far considered a single substation, along with its connections to the wide area network. However, a real power transmission system contains a large number of substations.

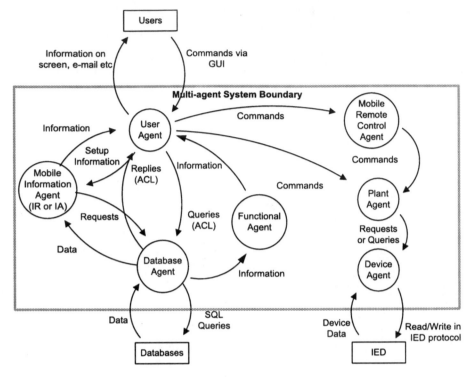

Fig. 3.9. Agent collaboration

It is envisaged that in the architecture described each substation would contain a multi-agent system consisting of a data acquisition system and information management system. This would produce a modular architecture consisting of individual substation modules and WAN modules. While possible, communications between substations would be limited, as many functions do not require direct inter-substation communications. Each substation should have its own agent platform and directory services, reducing the amount of data transmission across the wide area network. However, it would be necessary to allow user agents, and other agents located on the wide area network, to have access to any/all substations.

3.4 Agents, Tasks and Interaction Protocols

Each task described in Section 3.2 is associated with a particular type of data and a particular interaction protocol, taken from the standard FIPA interaction protocols. Tables 3.1 and 3.2 summarise these associations. Table 3.1 considers each task as a transformation from input data to output data, and defines the types of data involved and the agents that perform the task. Table 3.2 considers the characteristics of the task (whether it is event-driven or performed on demand by some user or system) and the protocols involved.

Table 3.1. Tasks, data types and agents (DB = database, DS = data storage)

Task	Input Data	Output Data	Agents
User Interaction			User Agent
- Queries	HMI input	ACL queries	
- Requests	HMI input	ACL queries	
- Online Display	Events or Status	HMI display	
Intervention	Requests	Requests	Plant Agent
Output D.I.	Required plant status	DAQ actions	Plant Agent
Automatic Control	Current plant status	Required plant status	Plant Agent
Input D.I.	DAQ Events/ Status	Plant Events/ Status	Plant Agent
Data Acquisition	Sensor data	Events/Status (DAQ system)	Device Agent
Data Storage	Data in ACL format and global ontology	Data in SQL statements and DB schema	DB Agent or DS Agent
Data Querying	Queries (ACL)	Responses (ACL)	DB Agent
Document Retrieval	Queries (ACL)	Binary Data	Document Agent
Document Storage	Binary Data in ACL	Binary Data in files	Document or Document Storage Agent
Information Gathering	ACL queries	Information in ACL	Mobile Agents
Analogue/Digital Output	New values (ACL)	DAQ Channel Values	Device Agent

The following sections describe how the individual tasks are carried out. It is assumed that agents have already located each other and therefore searches of the DF to locate appropriate agents are not included in the descriptions.

User Interaction

User interaction (Figure 3.10) is carried out by the user agent, which takes commands and queries from the user (via the HMI) and translates them into appropriate ACL messages for transmission to other agents.

Table 3.2. Tasks and interaction protocols

Task	Characteristics	Protocols
User Interaction		
-Queries	On-demand	FIPA Query
-Requests	Request/Reply	FIPA Request
-Online Display	Event-driven	FIPA Subscribe
Intervention	Request/Reply	FIPA Request
Output D.I.	On request	FIPA Request
Automatic Control	Event-driven	FIPA Subscribe (input)
		FIPA Request (output)
Input D.I.	Event-driven	FIPA Subscribe
Data Acquisition	Event-driven	Device-dependent(input)
		FIPA Subscribe (output)
Data Storage	Event-driven	FIPA Subscribe (data gathering)
Data Querying	On-Demand	FIPA Query
Document Retrieval	On-Demand	FIPA Request, FIPA Query
Document Storage	On-Demand	FIPA Request
Information Gathering	On-Demand	FIPA Query, (FIPA Request)
Output (A or D)	On-Demand	FIPA Request

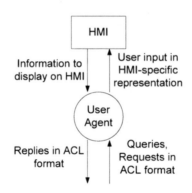

Fig. 3.10. User interaction

Intervention (Including Output Data Interpretation)

Intervention (Figure 3.11) is performed by the user agent and plant agent. Upon receiving a command from the user via the graphical interface, the user agent searches the DF to locate a plant agent capable of carrying out that command. It then forwards the command to the appropriate agent as a FIPA ACL *request* message. The plant agent generates a sequence of read or write operations to carry out in order to fulfil the request, and transmits these to the device agent.

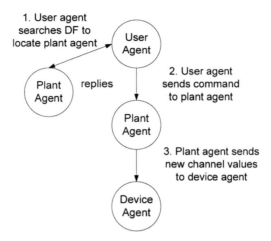

Fig. 3.11. Intervention (including output data interpretation)

Data Acquisition

Data acquisition (Figure 3.12) is performed by the device agents. The device agent must first establish a connection to its IED or other device, and acquire the values of its channels either by polling or by an event-driven method.

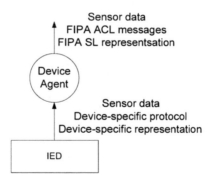

Fig. 3.12. Data acquisition

Input Data Interpretation

Input data interpretation (Figure 3.13) is performed by the plant agent, in cooperation with the device agents. The device agent must first register the channels that it is monitoring in the DF. Plant agents may then search the

DF to locate device agents monitoring those channels that are connected to their items of plant. Once a plant agent has found the relevant device agent or agents, it then establishes subscriptions with these agents in order to be notified whenever the value of a relevant channel changes. After these subscriptions are established, the device agent must notify the plant agent whenever the value of a subscribed channel changes. This notification will be a FIPA ACL message with FIPA SL content, giving the name of the channel and its current value. The device agent then uses this information to derive the state of its item of plant. The mechanism used to do this is described in Section 4.5.

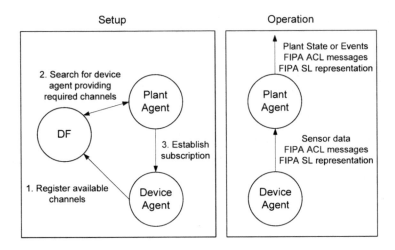

Fig. 3.13. Input data interpretation

Automatic Control (Including Output Data Interpretation)

Control (Figure 3.14) is performed by the plant agent. New channel values are passed to the plant agent by the device agent. The plant agent then performs input data interpretation on these values to generate the state of the item of plant being controlled. The agent then uses this state and the desired state to generate a set of outputs to be written to the plant, which are sent to the device agent.

Data Storage

Data storage (Figure 3.15) is performed either by the database agent or by a specialised data storage agent (this topic is discussed in Section 3.3.3). The

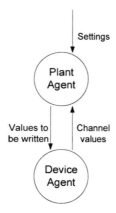

Fig. 3.14. Automatic control

database agent must first establish a subscription with appropriate provider
agents in order to be notified of new data. As events arrive, it generates SQL
update statements from these events, and enters the new information into the
database.

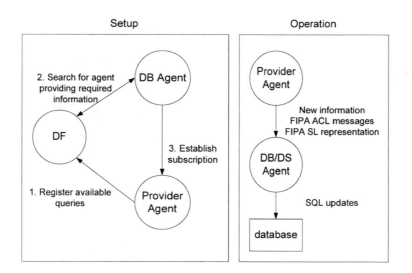

Fig. 3.15. Data storage

Data Querying

Querying (Figure 3.16) is performed by the database agent. A client agent (usually the user interface agent or a mobile agent) generates a query in FIPA ACL/FIPA SL format. The database agent then maps that query into an SQL query which is sent to the database. The results of the query are converted into FIPA ACL/FIPA SL and returned to the client. Integration of data from multiple databases may also be performed at this stage. If a querying agent determines (using the directory facilitator) that multiple agents have information relevant to the query, it may query both of these agents and then merge the results retrieved. Because all database agents convert data into the global ontology, this should be a relatively simple process in most circumstances. A similar procedure may also be used for answering complex queries, for example, to retrieve all transformers exceeding their load rating. In this case, the querying agent might retrieve the list of transformers and load ratings from the static database, and then search the data logging database to determine the maximum load of each one.

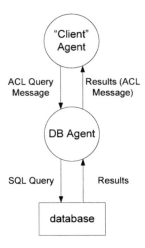

Fig. 3.16. Querying

Document Retrieval

The document retrieval process (Figure 3.17) consists of two operations. First, the client agent must determine which documents are relevant to a particular query. Then it may retrieve some or all of these documents.

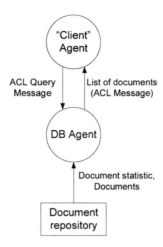

Fig. 3.17. Document retrieval

Document Storage

A document may be added to a document repository either directly or via a document agent. To add a document directly, its file is placed in the filesystem of the document repository. To add a document via a document agent, it may be sent to the agent as the content of an ACL message, using Base 64 encoding to convert a binary document into a text-based representation suitable for transmission.

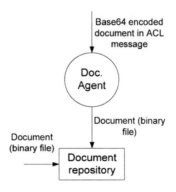

Fig. 3.18. Document storage

Information Gathering

The information gathering process performed by mobile agents is described in Chapter 5.

Analogue/Digital Output

Output to devices is performed by the device agent, which translates between the ACL representation used by other agents and the device-specific representation.

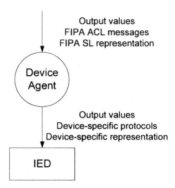

Fig. 3.19. Analogue/Digital output

3.5 Data and Knowledge

In order to permit exchange of knowledge between agents, it is important to have a standard representation of the knowledge available in a substation. The representation chosen in this system is based on first-order logic, and in particular on the form used by FIPA SL [36]. This section first discusses the different types of knowledge available in the power system, and then considers its machine-understandable representation.

3.5.1 Available Data and Knowledge

In a substation automation system, the data available largely consists of status information, "streamed" data, configuration information and other fixed data sources such as documentation.[3]

---

[3] These categories are partly based on those described by the IEC 61850 standard [91, p41]

Status and Event Information

consists of both state information, which gives the current state of some part of the substation plant, and event information, which describes changes in the state of the plant. Often, a data point gives the value of some input, which corresponds to a property of an item of plant. For example, "the low-voltage current of transformer 1 is 137.5 kV". Events are usually associated with a time tag.

When real-time information is stored for processing at a later time it is referred to as *historical* data. Historical data includes, for example, logs of events and stored waveforms.

Streamed Data

If a data point is monitored in real-time by another process, the waveform data will be transferred, a single item at a time, from one process to another. This is referred to as "streaming", and may be used in a variety of real-time monitoring applications.

Configuration Information

Configuration information consists of information about the configuration of the substation and about the configuration of the automation system itself. For example, topology information regarding the connections between different items of plant is included in this category.

Non-machine-understandable Data Sources

make up a further category of information. Currently, non-machine-understandable data sources include documentation and images. This data differs from the real-time and configuration information in that it is not immediately usable by a computer system in a reasoning process. However, it is possible that various processes, for example text mining or image recognition, might be able to derive machine-understandable data from non-machine-understandable data.

Derived Knowledge

As well as the "raw data" discussed above, which is considered to be the inputs to the automation system, the system may itself be capable of deriving additional knowledge. For example, National Grid Technical Specification (NGTS) 2.7 states that a substation control system is required to "maintain a running total of operations for each of the primary plant items" [116]. Another example of derived knowledge is the output of a condition monitoring system which uses real-time information (*e.g.* temperature data) to determine whether or not a transformer is in a good operating condition.

3.5.2 Knowledge Representation

The representation of each of the types of knowledge described above in first-order logic, and hence in the FIPA Semantic Language (SL), is now considered. The notation from Russell and Norvig [117], with mathematical expressions and sets denoted as in [117], and not FIPA SL, is used here. In this notation, object and relation names start with capital letters, and variable names with small letters. Where frames or identifying expressions are used the FIPA SL notation is used instead, as these are not included in the standard notation. FIPA SL is more difficult to read and produces longer expressions than standard logical notation. However, conversion to the FIPA SL notation from the notation used here is relatively simple, as there are only syntactic and not semantic differences. The main difference is that FIPA SL uses an infix notation in which the entire expression is enclosed in parentheses. Also, variable names start with a question mark, and all other names may start with either small or capital letters. For example, "if x is a transformer, then x is an item of plant" could be written in first-order logic as follows:

$$\forall x \, Transformer(x) \Rightarrow Plant(x)$$

and represented in FIPA SL as:

```
(forall ?x (implies (transformer ?x) (plant ?x)))
```

In the above expression, the $\forall$ symbol (for all) means that the statement applies to all possible values of the variable x. The $\Rightarrow$ symbol (implies) is used to state that if the statement on the left ($Transformer(x)$) is true, then the statement on the right ($Plant(x)$) is also true. FIPA SL also includes modal extensions in the form of the operators B (believes), U (uncertain), I(intends) and PG(is a goal of). Each of these operators takes two arguments: an agent and a logical statement. For example:

```
(B agent1 (status x110 open))
```

means that the agent "agent1" believes the statement "(status x110 open)", which might mean "the circuit breaker X110 has status open". The use of modal logics in reasoning about agents is discussed in [118].

Representation of Status information

If we assume that a single data point of status information gives the value of some property of a plant item (or other item of equipment) at a particular time, it is possible to represent a data point by a binary relationship, or subject-predicate-object triple. For example, suppose that we have the statement "The low-voltage current of transformer SGT1 is 12.3 volts". In first-order logic, this could be represented by:

$$LvCurrent(Sgt1, 12.3)$$

Representation of Events

An event must be able to represent the fact that at a given time, some change occurred in the status of the plant, or some action was performed. Due to the requirements of the various processes in the power system, the time represented in such an event must be explicit, rather than defined using operators such as "PAST" or "NEXT" as may be done in temporal logic [119]. We adopt an event representation based on the event calculus representation of time described in [117], Chapter 8. A formal semantics and logic for such a representation is established by Shoham in [120]. An event is represented by the T (or $TRUE$) "predicate", with two arguments: the event itself, and a timestamp, the format of which is discussed below, giving the time at which that event occurred.

For example, the event "at 12:30 pm on 03/02/2003, circuit breaker X1 was opened" might be represented by:

$$T(Opened(X1), 20030203T123059000)$$

Events may also involve changes in the value of a quantity, for example, suppose the low-voltage current of the transformer SGT1 was 239.39 volts at a given time:

$$T(LvCurrent(Sgt1, 239.39), 20030203T123059000)$$

Russell [117] states that an argument to the T predicate, such as $LvCurrent$ (Sgt1,239.39) in the example above, must not be treated as a sentence in predicate calculus (the T predicate, as with other predicates in first-order logic, may take only terms as its arguments, and the statement $LvCurrent(Sgt1, 239.39)$ is a sentence rather than a term). Therefore, $LvCurrent(Sgt1, 239.39)$ must be treated as a function, generating an event as its result. However, Shoham [120] disagrees with this "reification" approach (treating a statement as a function) because he believes that its semantics are unclear, and proposes that a different semantics based on modal logic be used (although both approaches share the same syntax and the T or $TRUE$ notation).

Several alternative syntactic representations of the event calculus may be found in the literature, such as the representation used by Kowalski and Sergot [121], which denotes each event by an identifier.

Timestamps

In the examples above, the timestamp was given using the FIPA SL [36] standard time representation. This time format consists of a four digit year, two digit month and two digit date, followed by the letter "T", followed by a two digit hour, two digit minute, two digit second and three digit milliseconds value. This is accurate (assuming absolute accuracy of the device that produced the timestamp) to within 1 millisecond, which is also the most commonly used accuracy for time tagging of substation events within the power industry [122]. However, for certain functions such as phase angle measurement [122, 123], more accurate time tagging is required. It might also be useful to represent the accuracy of a particular timestamp, for example, to state that the timestamp of a particular event was accurate to +/- 1 ms. For this purpose, a timestamp object, represented as a FIPA SL functional term with parameters, is proposed. This has two components: the timestamp itself, consisting of year, month, day, hour, minute, second, millisecond components, and its accuracy. Smaller time components than 1 millisecond will be represented by a floating point millisecond value, *e.g.* 123.45 milliseconds for 123 milliseconds and 450 microseconds.

An example of a timestamp frame is shown below:

```
(timestamp
     :time (time :year 2002 :month 12 :day 10
                 :hours 10 :minutes 5 :seconds 2
                 :milliseconds 20.34
           )
     :accuracy (time :milliseconds 10))
```

However, as can be seen from the example, this produces a very large textual representation which might reduce performance in transferring data. A possible abbreviation would be to use the standard FIPA time representation for the time and an integer value (representing milliseconds) for the accuracy, adding decimal places to the time representation as follows:

```
(timestamp
     :time 20020204T123456789.500
     :accuracy 0.1)
```

The drawback of this is that the adding of decimal places to the timestamp does not comply with the FIPA standard.

Representation of Non-machine-understandable (Binary) Data

It is assumed that non-machine-understandable data, such as documents and images, is stored on disk in a binary format. However, in order to transfer and display this information, it is necessary to represent some basic facts regarding it. This information is known as *metadata* (data about data). A standard set of terms, the Dublin Core vocabulary [124], are available for use in the representation of metadata and are widely used on the Internet, and are adopted for the system described here.

Representation of Derived Knowledge

Unlike event data, which can be put into a single, fixed format, it is not possible to do so in general for derived knowledge, as many different types of knowledge may be provided. For example, "transformer sgt1 is in good condition" can be considered as a binary relationship giving the value (good) of a property (condition) of a particular object (transformer sgt1), and might be represented by $Condition(Sgt1, Good)$. "The oil input temperature of transformer Y is between 30 and 40 degrees" is similar, but the value of the property is an interval rather than a single value. This might be represented by the statement "There is some x such that the input temperature of sgt1 is x and x is less than 40 and x is more than 30":

$$\exists x\, InputTemperature(Sgt1, x) \wedge (x < 40) \wedge (x > 30)$$

However, problems might exist when reasoning with this representation or storing it in a database, as many programming languages do not allow interval values for variables.

Finally, there are predictions made by condition monitoring systems, for example "if the ambient temperature does not rise then transformer Y will not exceed its operating requirements if the load increases by 50%" (this scenario is based on the system described in [125]). The representation of this statement is outside what may be represented in standard first-order logic. Roughly speaking, it might be interpreted as "Agent 1 (the agent making the prediction) believes that if the load on transformer Y is equal to Z and nothing else changes from its current situation, then transformer Y's condition will be good" (this assumes that transformer Y's condition is already good). The "Agent 1 believes" part of the statement might be represented using a modal logic of belief, as supported by FIPA SL. For example (using B to denote the "believes" predicate), it is possible to state "Agent 1 believes that the load on transformer SGT1 being l does not imply that the condition of SGT1 is bad":

$$B(Agent1, \neg(Load(Sgt1, l) \Rightarrow Condition(Sgt1, Bad)))$$

This is not sufficient to conclude that if the load on sgt1 is l then the condition of sgt1 will be good. In order to do this, the other conditions affecting sgt1's condition must also be stated (*e.g.* ambient temperature).

A simpler representation of a prediction might be to represent it as the result of a "predict" action taken by some agent. For example, an agent might predict that the hotspot temperature of a transformer at time <time> under load <load> would be 92.5 degrees centigrade:

```
(result
    (action
        (agent-identifier :name prediction-agent)
        (predict-hotspot-temperature
            :time <time>
            :load <load>
        ))
        92.5)
```

3.5.3 Ontologies

In addition to a common knowledge representation, a common *ontology* is also
required for interoperability between agents. The ontology provides a defined
set of terms to be used for communication, and hence eliminates problems
caused by the use of different terms for the same physical object or quantity, for example, one agent referring to the temperature of a transformer
as "temperature" and another as "temp". For example, we may define that
"low-voltage current" of a transformer is to be represented by a binary relation named *lv-current*. Once such an ontology is agreed upon, it is possible
to use the multi-agent system to integrate data from heterogeneous sources,
by converting the heterogeneous knowledge representations into that of the
global ontology [126].

The ontology used in this system must consist of at least the following
sub-ontologies (Figure 3.20), implementations of which in a prototype system
are described in Chapter 6.

- The FIPA meta ontology, and in particular, that agents understand the
  slot-of, template-slot-of, subclass-of, instance-of relations, and where appropriate their inverse relations. This allows agents to use the concept
  of inheritance to derive facts about objects from their classes and facts
  about classes from their superclasses. It also allows, for example, the user
  interface agent to determine the available properties of an item of plant
  (using *slot-of*), which permits it to generate a list of properties that may
  be queried or used in a data analysis operation.
- An ontology describing substation plant, including the different classes of
  plant (circuit breaker, transformer *etc.*) and the properties of those classes
  (for example, the tap position of a transformer). This ontology is used
  for the exchange of data and events regarding the substation. This is an
  example of a *domain ontology* (as defined in [126]).
- An ontology describing the components of the data acquisition and automation system. This permits agents to describe components of the data
  acquisition system, and perform operations such as system configuration.
  As with the plant ontology, this is a domain ontology.
- An ontology for information management, including concepts such as information resources, queries, and the relevance of an information resource
  to a query. This is a *task ontology* (as defined in [126]).

Different agents use different subsets of the global ontology. For example:

- The *plant* class, which is the superclass of all substation automation equipment in the substation plant ontology, is shared by the user interface agent
  and the static database. The user agent queries the static database for
  subclasses of *plant* (to generate a list of types of substation plant), and
  then generates a list of instances of a chosen subclass using the *instance-of*
  relation. However, when programming the user interface agent, it is unnec-
  essary to include any more of the substation plant ontology, as this may

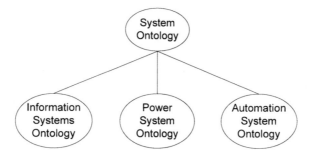

Fig. 3.20. Hierarchy of ontologies

be obtained from the database agents. This lack of implicit knowledge of the substation domain embedded in the user interface agent could allow it to be used in a variety of different (industrial automation) domains, providing that a graphical user interface for the alternative domain was available.

3.6 Agent Platform Implementation

An agent platform is a software system providing a set of standard services (*e.g.* AMS, DF and ACC) to perform agent lifecycle management, communications and service discovery [77]. All agents are associated with a particular agent platform. Many agent platforms also provide a set of libraries in some programming language or languages (usually Java) which simplify the task of the agent programmer. The FIPA specifications define the services that must be provided by an agent platform, and how these services should be accessed from outside the platform. However, the internal architecture of the platform is dependent on the implementation. There are several agent platforms available which conform to the FIPA specifications.

For the initial prototype stage of the project, a mobile agent platform was implemented based on the FIPA [49] specifications. However, for the main prototype described in Chapter 6, the JADE [71] platform, which is a Java-based agent platform compliant with the FIPA specifications, was chosen, because at the time this choice was made it was the most commonly used FIPA platform, and was the only freely available FIPA platform with mobile agent support. However, the architecture itself is not dependent on any particular platform. This section examines some of the implementation choices that would have to be made in the development of a system based on this architecture.

3.6.1 Standard FIPA Platform

There are a number of standard FIPA platforms available [127], and one of these could be used for the development of the system, as was done for the prototype described in Chapter 5.

The main advantage of using such a platform is the reduction in the amount of implementation effort required to implement the architecture through the use of an off-the-shelf agent platform. Also, the platform would be able to interoperate with other FIPA systems if that was required.

The major disadvantages of using a FIPA platform are the fact that no FIPA platform provides inter-platform mobile agent support (JADE supports intra-platform agent mobility), and this would have to be added, if required, either by modifying the platform or by implementing a simple mobile agent system which would interoperate with this platform through the FIPA IIOP interface and be used solely for hosting mobile agents. Also, most current FIPA platforms are based on Java and CORBA®. This means that their use for real-time applications, or for hosting agents on substation devices, is restricted, although there are Personal Java^{TM}or Java^{TM}2 Micro Edition compatible platforms available.

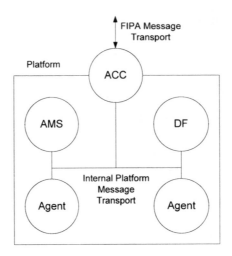

Fig. 3.21. Abstract FIPA agent platform (simplified)

Figure 3.21 shows the abstract architecture of a FIPA agent platform. The standard services (AMS, ACC and DF) provide lifecycle management, communications and messaging facilities and service lookup respectively. All agents, including these, communicate via the Internal Platform Message Transport (IPMT). For communications with agents on other platforms, the ACC communicates with a remote platform ACC via the FIPA Message Transport service. However, when considering implementation issues it is simpler to discuss a particular instantiation of this architecture. Figure 3.22 shows the architecture of the JADE agent platform, which is FIPA-compliant and is developed by Telecom Italia Laboratories and the University of Parma [71]. JADE's architecture is based on a modular structure, in which a platform is split into a number of containers, which may run on different machines and communicate via Java^{TM} RMI. Communication between agents on the same container uses Java^{TM} method calls.

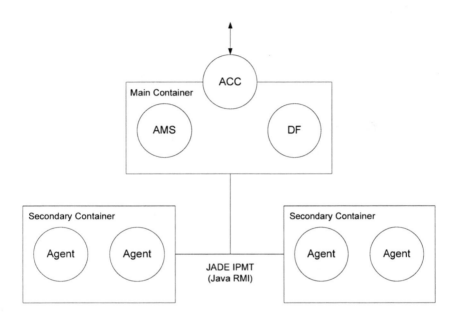

Fig. 3.22. Java agent development environment architecture

3.6.2 Jini^{TM}-based Platform

Jini^{TM} is a system for service management and discovery developed by Sun Microsystems and based on Java^{TM} and Java^{TM} RMI. It would be possible to implement an agent platform based on Jini^{TM}. In fact, there are several such platforms in existence, for example, the Ronin platform [128], and the platform used by the PABADIS project [85]. Such a Jini^{TM}-based platform might be implemented as follows:

- A standard Jini^{TM} service is defined for an agent, which has a single method, allowing agents to send messages to that agent. This is based on the IIOP interface specified by FIPA

- A gateway provides FIPA services and allows the platform to interoperate with FIPA systems.

A design for this platform is shown in Figure 3.23.

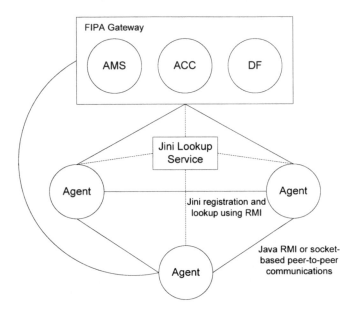

Fig. 3.23. Jini^{TM}-based agent platform

The main interesting feature of the implementation described here would be its highly decentralised nature, in which there are very few essential platform services. The only requirement would be the Jini^{TM}lookup server. Other services such as brokers and the FIPA gateway would themselves be agents, rather than components of the platform. The Jini^{TM}architecture provides high reliability compared to other agent platforms, as it includes features such as leasing, which prevents references from remaining in the lookup service after a service has terminated. It is claimed that leasing allows Jini^{TM}to "self-heal" a distributed system [129].

The problem with creating a Jini^{TM}-based implementation is that a substantial amount of implementation work would be required to make such a platform FIPA compatible. It would be necessary to provide "wrappers" for the Jini^{TM}lookup service to allow it to be used as a FIPA Directory Facilitator, and a service would have to be provided, either by each agent individually or by a gateway agent, to permit agents to send and receive FIPA ACL messages.

3.6.3 UDP-based Platform

A very simple agent platform, with limited infrastructure, would be one based on the use of the User Datagram Protocol (UDP) to send messages between agents. Each agent would listen for messages on a UDP port, and agents would also be able to listen for broadcast messages on a broadcast port. This allows for both peer-to-peer and broadcast messaging. In the context of this infrastructure, a "platform" would be a single subnet of the network, in which all agents were within the range of each others' broadcast messages.

The structure of the UDP-based platform is identical to that of the Jini^{TM}-based platform, except that the protocol used both for inter-agent and agent to lookup server communications is UDP rather than Java^{TM} RMI. The central lookup server would be used to enable agents outside the platform, who were not capable of sending broadcast messages into the network, to locate server agents inside the platform. A FIPA gateway would also need to be provided, which would also act as a message router and lookup service for inter-platform communications.

Problems might be caused when implementing this platform by the lack of ability of certain hubs and switches to transmit broadcast messages. This would have to be examined further were this platform to be implemented.

The main advantage of the UDP-based platform is that the UDP protocol provides for extremely fast message transmission. Therefore this platform might be better suited for the implementation of agents required to work under timing constraints, such as device agents. Also, because the agents do not execute on a platform server, they may be written using any programming language. However, this can also be achieved to a lesser extent when using a FIPA platform, as agents can interoperate with and register on the platform using the IIOP interface.

The drawbacks with this platform, as with the Jini^{TM} platform, mostly concern the amount of implementation work required. Also, the platform may not be as robust as the Jini^{TM} platform because UDP does not provide reliability features, although these could be implemented at platform level.

3.6.4 Combined FIPA and UDP-based Platform

A possible solution to the real-time and limited device difficulties of a FIPA platform and the scalability problems and implementation overhead of a UDP-based platform would be to combine the two. This combined platform might have a similar conceptual architecture to that of the Lightweight and Extensible Agent Platform (LEAP) system [130], which extends the JADE FIPA platform by providing a "container" based on the Java^{TM} 2 Micro Edition, which hosts JADE agents and communicates with a host platform using a proprietary socket-based protocol. However, the platform proposed here is slightly different, in that it consists of a FIPA-based platform along with a gateway agent, which allows any number of independent agents to join the system. These agents are implemented as described for the stand-alone UDP-based platform.

The function of the gateway is to translate messages from the format used by the UDP-based agents to FIPA message objects as defined by the FIPA platform, and to pass them to the FIPA platform's Agent Communications Channel (ACC) agent for handling. This means that all platform services for the UDP-based platform can be provided by the FIPA platform, but the UDP-based agents retain the ability to communicate with each other in a peer-to-peer manner, without using the platform, when required.

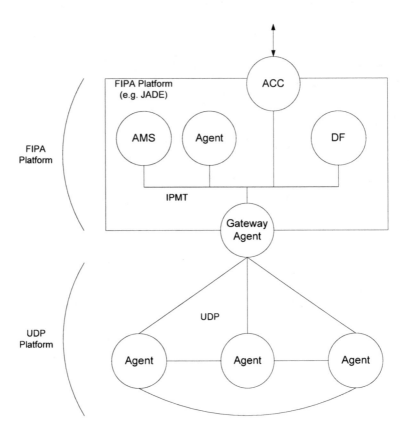

Fig. 3.24. Combined platform

3.7 Summary

This chapter has described a generic multi-agent software architecture for power system automation systems. The agent architecture is derived from the structure of the power system and data acquisition system. The device agents provide a view of the system based on the monitoring and control system, which consists of data acquisition devices. Meanwhile, the plant agents provide a representation based on the substation plant, which is convenient for the implementation of distributed control, and also for acquiring information about a specific item of plant. The use of one agent for each device or plant item provides a highly distributed architecture, and a natural representation of the physical system being controlled and monitored.

The representation of power system knowledge within the multi-agent system and the methods of communication between agents are also discussed. The system uses the logic-based language FIPA SL to encode events, historical data and commands. A basic ontology of automation systems and power systems is provided for this purpose. Agents communicate using various FIPA standard protocols. The FIPA subscribe protocol is used for regular monitoring, the FIPA query protocol for database and document querying, and the FIPA request protocol for the transmission of commands. The use of these three protocols permits a wide variety of interaction styles and allows many different tasks to be implemented.

Finally, this chapter has investigated the implementation of the agent platform. This is an important aspect of the system as it provides the basis for all agent execution and communications. Various possible platforms were described, including standard FIPA platforms and platforms based on Jini^{TM} and UDP. For the purposes of the prototype, the FIPA platform was chosen, as it was the most convenient and provided a simple method of implementation. However, for a full system, a hybrid UDP and FIPA platform might be more suitable, as the use of UDP provides for higher-speed communications, which are important in certain industrial applications.

The overall architecture having been considered in this chapter, the next chapter discusses the design and implementation of the individual static agents. The tasks that each agent must perform are used to derive the required capabilities, sensors and knowledge of the agent. Where appropriate, issues regarding agent implementation are also described.

4

Static Components of Architecture

This chapter describes the individual software agents that make up the architecture described in Chapter 3. All agents are based around the same basic architecture, shown in Figures 4.1 and 4.2. This is derived from standard descriptions of agents such as those in [8, 38, 117], and also draws on the BDI architecture, described in Section 2. Each agent consists of sensors, which allow the agent to perceive its environment, a knowledge base containing the agent's beliefs and goals, and effectors, which allow the agent to take actions. A reasoning engine takes inputs from the sensors and knowledge base, and determines the actions to carry out. This may be done in different ways, for example, in a BDI agent plans stored as part of the agent's knowledge are used. It is intended that the architecture used should allow different forms of reasoning to be used and does not commit the agent to using any specific methodology. However, individual agents may be, for example, BDI agents, in which case the reasoning engine of the agent would be a BDI interpreter. By creating agents with different reasoning processes, knowledge, sensors and effectors, different tasks may be carried out.

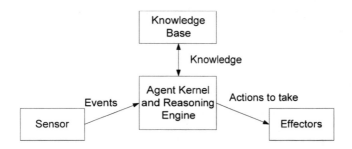

Fig. 4.1. Agent architecture

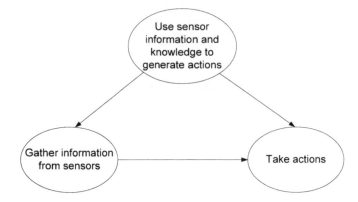

Fig. 4.2. Agent control loop

The agent specifications provided in this chapter state the knowledge, capabilities, sensors, effectors and interaction protocols of each agent. The sensors provide input to the agent, the capabilities (effectors) allow the agent to perform actions and the interaction protocols provide the means for the agent to engage in standard conversations and therefore to collaborate with other agents.

In addition to this, the agent specifications also detail which components of the system ontologies (described in Section 3.5.3) are used in the conversations of each agent. It is assumed that all agents use the *FIPA Agent Management* ontology (for registering with and searching the DF to enable agents providing a particular service to be located).

As well as the specifications of the individual agents, this chapter also discusses some of the implementation and knowledge representation issues involved with certain types of agent. These include the implementation of distributed database querying in the database agents, the use of BDI mental components to represent an agent, and the representation of mapping rules for translating sensor data into a plant state representation in the plant agents.

4.1 Database Agents

4.1.1 Description

The database agents provide the FIPA ACL-based access to a database. Database agents of this type have previously been used in a wide range of information management applications and architectures, for example [82]. The structure of the agent is shown in Figure 4.3. In order to fulfil its task, the database agent must be able to insert and retrieve information to and from the database and to convert this information to FIPA ACL and into the system's global data model (ontology). This conversion is performed by the agent's reasoning engine using mapping rules stored in the knowledge base of the agent, which specify, for example, that a particular property of an item of plant corresponds to a given column in a database table. The format of these depends on the particular implementation of the agent used. The agent's communications sensor and effectors allow it to send and receive information to other agents.

Unlike most of the other agents in the system, the database agent does not implement the *fipa-subscribe* protocol, which means that it is unable to provide subscription-based access to a database, in which subscribing agents would be notified of new data as it arrived. While it would be possible for it to do so, the only method of implementation available with most current database management systems would be to "poll" the database at regular intervals to determine if new knowledge had been added. There are two exceptions to this statement:

1. If the database agent is the only entity capable of inserting information into the database, it will be able to provide subscription functionality as the new data will be passed through the database agent.
2. If an "active database" [131] is used, it might be possible to add a rule to the database requiring it to generate an event on the arrival of particular data and pass that event to the database agent by some means (possibly specific to the particular database).

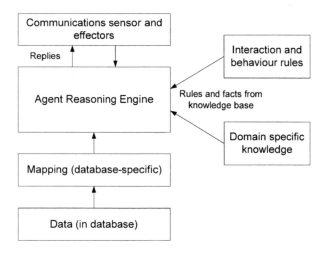

Fig. 4.3. Structure of database agent

4.1.2 Agent Specification

This specification defines the knowledge, capabilities, sensors, interaction protocols and ontological knowledge required by the database agent. Using this information, it is possible to modify a generic agent to act as a database agent.

- Knowledge
  - Transformation rules from database schema to global schema.
  - Database configuration (Java Database Connectivity driver, username, password).
- Capabilities
  - Query database
  - Append data to database[1]
- Sensors
  - No explicit sensors (apart from communications). This agent exists only in a software environment, and has no connection to the outside world except through other agents.
- Interaction Protocols
  - FIPA query (responder)−used by other agents to query the database
  - FIPA request (responder)−if data is to be added to database on the initiative of another agent rather than via a subscription.
  - FIPA subscribe (initiator)−this allows the database agent to establish a subscription with a data providing agent (*e.g.* a plant agent) so that new events are transferred into the database as they occur.
- Ontology
  - The agent converses using components of the global ontology relevant to the knowledge contained in its own database. This will differ for different database agents and is implicit in the mapping rules of the agent.

4.1.3 Agent Implementation

It is possible to implement the reasoning of a database agent in a number of different ways. This is particularly relevant to the need to integrate information from multiple databases. Different implementations allow this to be performed in different ways. Also, certain implementations may provide additional functionality, which might be useful in a particular application. Here three alternative implementations are considered, a simple agent, which only acts in direct response to queries from other agents, a simple collaborative agent, which is capable of querying other agents, and a BDI agent.

---

[1] In some applications, for example the substation information management application described later in this book, the database agent is not able to alter the database, but can only query it.

Simple Implementation

In a simple implementation, the agent implements the FIPA query protocol, but does not have any autonomy: it will not provide information unless it is asked for it. The agent must wait until it receives either a *query-if* or *query-ref* message. It then uses the content of that message to generate an SQL query to the database.

Simple Collaborative Implementation

This implementation, based on previous work such as [108] and on the techniques used by distributed Prolog implementations, extends the capabilities of the database agent by allowing it to communicate with other database agents and integrate information. The agent uses the DF to determine queries that may be posed to other agents, represented as referential expressions (as defined in the FIPA SL specification [36]). For example, an agent might advertise that it has knowledge about the status of circuit breaker "h13" using the referential expression "(iota ?a (status h13 ?a))". When performing a query, a database agent will retrieve information from its own database where possible, and if not possible, will forward the query, or the part of the query that it is unable to answer, to another agent that is capable of answering it.

This might employ a distributed backtracking procedure, as has been implemented in distributed Prolog systems such as those discussed by [132]. Alternatively, it might be better to implement this collaborative capability as a separate broker agent, as is done in the Infomaster^{TM} and RETSINA systems [80, 82]. This would reduce the load on the database agent and reduce the need for parallel query processing.

BDI Implementation

In a BDI implementation, we give the agent explicit goals regarding the information it is to provide to other agents, as well as allowing it to reason about other agents' information needs. This can either be used to extend the behaviour of the agent, for example to provide the agent with the ability to remember past queries as proposed by [133], or just to provide an alternative conceptualisation. In the BDI implementation, whenever a query with content p arrives from another agent a a goal is added to achieve "a knows p". Thereafter, unless the database agent comes to believe that a already knows p (perhaps because the database agent has already informed a of p), or that it is impossible that a will come to know p, the database agent will attempt to make a know p. This allows for two responses from the database agent: either to notify a immediately of the answer (if it is available), or at a subsequent time, as the information becomes available, to notify a. The agent may also decide to perform some action that would lead to it knowing p, perhaps querying another agent.

Possible alternative methods to achieve this "persistent query" behaviour are the use of the subscribe interaction protocol (although this results in the agent being notified whenever a value changes rather than just the first time it becomes available) or the *request-when* communicative act. However, these methods require the querying agent to explicitly specify that it wishes to be informed of something at a later time.

Choice of Implementation

In the prototype system, it was decided to use the simple collaborative implementation. This permitted agents to query each other for information that was not present in their particular database. For example, the static database agent was able to retrieve the properties of its items of plant using the ontology database agent. The BDI implementation was not used because of the amount of time required to implement it and because the functionality provided was not required. However, it would be considered as a possible enhancement for the system in the future. The simple implementation could also be used, but in this case a broker agent would be required to perform the information integration, or alternatively this functionality could be added to the user agent. However, this would increase the complexity of the system either by adding another agent or by making the user agent more complex.

4.2 Document Agents

4.2.1 Description

The document agent is responsible for the management of documents stored in a particular location, such as a directory of a filesystem. The agent has two main responsibilities: to ensure that it stores up-to-date statistics regarding its document collection, and to carry out queries on behalf of other agents. The structure of the document agent is shown in Figure 4.4. This is similar to that of the database agent. However, the document agent requires a sensor to notify it when new documents are added to the repository, and the ability to count the words in a document and generate metadata for use by an information retrieval algorithm. This algorithm can then be used to determine the relevance of a document to a particular query. In the prototype system, the information retrieval algorithm used is the standard Term Frequency/Inverse Document Frequency ranking algorithm [134]. The reasoning engine used by the agent is relatively simple. When queried it passes the query to the document ranking algorithm, which generates a set of matching documents. These are then returned to the querying agent. Agents can also use the document agent to retrieve the full text of a document.

A possible extension of the document agent would be to enable it to forward new documents as they arrive to interested parties, based on past queries.

This would be relatively simple, as the agent already has a sensor notifying it when new documents arrive. The required addition would be a list of past queries from users. The agent could then run a user's past queries against a new document to determine whether or not it is relevant to any of them, and if so, forward the document to the user. This is similar to the basic methodology used by Selective Dissemination of Information (SDI) systems, as described in [135].

As with the database agent, the document agent might be conceptualised as a BDI system. The definition and implementation would be similar to that for the database agent.

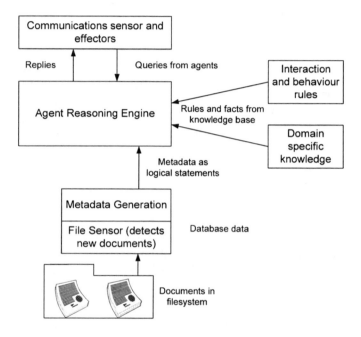

Fig. 4.4. Structure of document agent

4.2.2 Agent Specification

The knowledge, capabilities and sensors of the document agent allow it to generate document statistics and retrieve documents relevant to a query. To allow other agents to query for relevant documents, the FIPA Query protocol is used. The FIPA Request protocol allows other agents to request the text of a document.

- Knowledge

- Document collection statistics: These statistics must provide sufficient information for the agent to perform queries against its document collection. For example, the TF/IDF model [134] requires term frequency and document frequency statistics.
- Capabilities
  - Convert document to plain text: In order for document statistics to be generated, the plain text (ASCII/Unicode) of a document must be extracted from the provided representation. This ability (or set of abilities) allows the agent to do this.
  - Generate document statistics: This ability must be able to generate the required document statistics for the particular retrieval methodology used from the plain text of a document.
  - Retrieve list of documents relevant to a query.
  - Retrieve document full text on request.
- Sensors
  - Detect addition and removal of documents in a specified location. This sensor regularly scans the document repository, and generates an event whenever a new document is added or removed. This event may then be processed by the agent, which should react by generating statistics for the document and updating its document collection statistics.
- Interaction Protocols
  - FIPA Query
  - FIPA Request
- Ontology
  - The agent converses with other agents using the *information management* ontology. This ontology contains terms to describe documents and other information sources and the relevancy of an information source to a query.

4.2.3 Document Agent Issues

The main problem with the document agent is that methodologies for integrating the results of searches on multiple document collections, particularly when statistical methods such as TF-IDF are used by the individual collections, are still being researched. For example, [136] discusses different methodologies for distributed information retrieval. The authors state that, since document ranking heuristics such as TF-IDF make use of collection-dependent statistics, the relevance of a document to a query produced by one collection would not be the same as that produced for the same document by a different collection. Therefore, the authors conclude that "central coordination is necessary for aggregation of the results", particularly where the collection sizes are small.

In the architecture described, this issue could be handled by having a central "document broker" agent, which would hold aggregate statistics from the individual collections as in [136]. However, this might compromise the autonomy of the individual document agents. Further research into this area is required to determine the optimal solution.

4.3 Ontology Agents

The ontology agent used in this architecture implements the FIPA Ontology Service described in [109]. In addition, it registers a number of predicates with the DF as a "query service" as described in Section 3.3. Because the FIPA Ontology Service specification specifies the behaviour of this agent, and no autonomous behaviour is required, a simple database agent architecture may be used for this agent. Therefore, the implementation and structure of this agent is the same as that of the database agent described in Section 4.1. However, a set of rules specifying the FIPA Meta Ontology are required to permit it to accurately answer queries. For example, rather than specify separate rules regarding how to extract *superclass-of* and *subclass-of* from the database, it is possible to define *subclass-of* in terms of *superclass-of*:

$$SubclassOf(a, b) \Rightarrow SuperclassOf(b, a)$$

This requires that the agent has a reasoning engine capable of processing either logical statements or IF/THEN rules.

4.4 Device Agents and Node Agents

4.4.1 Description

A device agent is responsible for the management of a specific data acquisition device. The tasks of the device agent include both input/output and device configuration. However, data interpretation is the responsibility of the plant agents.

A node agent is responsible for managing a data acquisition "node". This corresponds to a single computer or IED, having a CPU and containing or controlling a number of data acquisition devices. Each node has a set of interfaces which enable it to communicate with other nodes and with the data acquisition system.

A device agent must maintain up-to-data knowledge of the current status of its device and the values of each input/output channel, which is achieved using the data acquisition sensor. It must also be able to provide this knowledge to other agents, in particular the plant agents, both via queries and subscriptions. Other agents must also be able to write to the output channels of the device using the FIPA-request protocol.

4.4.2 Agent Specification

The knowledge of a device agent relates to the device that it manages, and its capabilities allow it to interact with that device and with other agents. The node agent is provided with information about its node and has the capability to perform configuration tasks.

Device Agent

- Knowledge
  - Type of device (to select data acquisition sensor implementation—however, this might be implicit in the configuration of the agent rather than explicit knowledge)
  - Current status of device
  - Current value of each channel of the device
  - Device configuration
- Capabilities
  - Set the value of any output channel of the device.
  - Change device configuration (*e.g.* set sampling rate)
  - Start data acquisition
  - Stop data acquisition
- Sensors
  - Data acquisition sensor to permit data to be acquired from the device. This sensor should use an appropriate communications facility to maintain knowledge of the current state of each (input or output) channel of the device. Whenever a channel's value changes, an appropriate event should be generated. While the communications protocol used by the sensor depends on the device, the format of the event generated should be device-independent. Therefore, this sensor performs a mapping from the device's protocol to a standard event format. In addition, all events must be timestamped. If a timestamp is not provided by the data acquisition system, it should be added by the device agent, adding the appropriate level of accuracy to allow for data acquisition delays.
- Interaction Protocols
  - FIPA query—used by other agents to retrieve data from the data acquisition system.
  - FIPA subscribe—used by other agents to establish a subscription to be notified whenever the value of a channel changes.
  - FIPA request—used to write new values to channels of the device.
- Ontology
  - The agent describes its device using terms taken from the automation system ontology. This contains terms describing automation devices and channels. It therefore has no knowledge of the application domain (*e.g.* power systems) and does not use this domain ontology.

Node Agent

- Knowledge
  - Communications interfaces belonging to the node, and their configuration.
  - A list of devices belonging to that node.
- Capabilities
  - Change node configuration (exact changes will depend on the individual node).
- Sensors
  - The node agent is capable of reading the configuration of its node.
- Interaction Protocols
  - FIPA query—used by other agents to retrieve the configuration of the node.
  - FIPA request—used to request configuration changes.
- Ontology
  - As for the device agent, the automation systems ontology is used.

4.5 Plant Agents

4.5.1 Description

A plant agent is responsible for the monitoring and control of a single item of plant. In a substation, the main items of plant include circuit breakers, disconnecters, transformers, busbars, capacitors and reactors. For each of these categories, a distinct agent must be created, with appropriate knowledge and capabilities to enable it to represent that plant effectively. The following section discusses the possible design of these agents, which is used in the current prototype. However, because an automatic control application has not yet been implemented it has not been possible to validate the automatic control function of these agents. Therefore, refinements to the design may prove necessary in the light of experience from such an implementation. Another further enhancement which is not included in the design is collaboration between control agents of different items of plant.

A plant agent must maintain up-to-date knowledge of the state of its item of plant. The most efficient way to do this for dynamic properties (*e.g.* voltage, current) is to establish subscriptions with the relevant device agents. In this way the state information will only be transmitted when a value changes. Providing that the value of a property changes less often than the polling interval, subscription-based monitoring of that property is more efficient than regular polling of the device agent using query messages.

4.5.2 Agent Specifications

Here we provide agent specifications for a generic plant agent. For specific plant agents, *e.g.* transformer agents and circuit breaker agents, the abilities and knowledge would be customised to the specific item of plant. For example, the abilities of a circuit breaker agent would allow it to open and close the circuit breaker.

- Knowledge
    - Mapping from plant properties to channels monitoring these properties (obtained from mapping database on agent startup).
    - Static information regarding plant (obtained from static database on agent startup).
    - Dynamic properties of plant, *e.g.* voltage, current, *etc.* (obtained from monitoring agents on regular or event-driven basis).
    - Possibly knowledge regarding agents controlling other items of plant (this is an item for further work as it relates to automatic control of the plant).
- Capabilities
    - Specific to the device:
        · Transformer agent: set tap position and activate/deactivate oil cooler.
        · Switchgear agent: open or close item of switchgear.
- Sensors
    - The agent has no direct connection to the plant, but obtains information by communicating with other agents. Therefore, there are no explicit sensors.
- Interaction Protocols
    - FIPA query (responder)−used to allow other agents to query the status of the plant.
    - FIPA query (initiator)−used to obtain information from device agents.
    - FIPA request (responder)−used to allow other agents to control the plant.
    - FIPA request (initiator)−used to forward control requests to device agents.
    - FIPA subscribe (responder)−used by other agents to request updates whenever the status of the circuit breaker changes.
    - FIPA subscribe (initiator)−used to obtain regular updates from device agents.
- Ontology
    - For information management purposes, the plant agent uses the *FIPA meta ontology* to determine the class to which its item of plant belongs, and hence its properties. It can then use this information to set up subscriptions with appropriate device agents, using the *value* relationship and *channel* class from the *automation system ontology*.

– For control purposes, a set of plant control rules must be written using the plant ontology. As far as information management is concerned the agent is largely capable of self-configuration as the mapping rules may be read from the mapping database. However, the agent must currently be explicitly configured to control a specific device as there is no database for control rules. Further work should investigate the possibility of sharing control rules between similar plant agents (*e.g.* transformer agents), in the context of a specific application.

4.5.3 Data Acquisition System/Plant Mappings

As described in Section 3.3.2, each property of an item of plant may be measured or controlled by one or more channels of the data acquisition system. However, it is not always the case that there is a one-to-one correspondence between plant properties and data acquisition system channels. Several different types of mapping are possible:

Numerical Mappings (Input or Output)

are those in which the value of the plant property is determined by a mathematical expression in which the only other variable is the value of a single channel of the data acquisition system. For example, representing the value of a plant property by l and the value of a corresponding channel by c, a relevant numerical mapping might be:

$$l = 128.9c + 2.5$$

These mappings may be found in traditional SCADA systems (for example, National Instruments' Supervisory Control Toolkit for LabVIEW^{TM} permits the specification of linear or square root based scaling between raw analogue input data and tag values).

Enumeration Mappings (Input or Output)

are those in which a set of values $\{c_1, c_2...c_n\}$ of some channel map to a corresponding set of values $\{l_1, l_2...l_m\}$ of a plant property. The mapping may be either 1:1 or many:1. For example, the input values $\{0.0, 0.4\}$ of an input channel on a data acquisition system may map to the status values $\{open, closed\}$ of a particular circuit breaker. In a many-to-one mapping, for example, all values between 0 and 0.2 might map to "open", and all values between 0.2 and 0.4 to "closed".

Complex Actions (Output)

represent sequences of actions that must be undertaken in order to change the value of a property. For example, to alter the tap position of a transformer in the substation simulator described in Chapter 6, it is necessary to write a 1 to the "tap up" or "tap down" channel, and then wait for a response on another channel before writing a 0 to "tap up" or "tap down". To alter the tap position by more than 1, it is necessary to repeat this procedure the appropriate number of times. Complex actions may be represented by a state machine or by a procedure in a computer programming language.

Combinations of the Above

It may also be possible to perform mappings which are a combination of the above types, for example, converting a value using a mathematical formula and then mapping it to an element of an enumeration.

Representation of Mappings in First-order Logic

It is necessary for the agent knowledge base to contain a representation of the mappings described above. These could be implicitly specified in the programming of a device agent or plant agent. However, it is intended that agents should be able to exchange knowledge about these mappings, for example, to allow a plant agent to retrieve the mappings corresponding to its item of plant from a device agent or database agent. Therefore, a representation of each type of mapping in first-order logic is required. Here, simple atomic names are used for channels and items of plant; the actual representation of these objects may be more complex.

Simple mappings

may be represented as a simple formula, for example, the following states that if the value of a channel "Lvc" is v, then the low-voltage current of a transformer "Sgt1" is equal to v multiplied by 128.9, added to 2.5.

$$\forall v \, Value(Lvc, v) \Rightarrow LvCurrent(Sgt1, (v * 128.9) + 2.5)$$

assuming that the + and * functions are defined for the particular agent, and that the *value* predicate gives the value of a particular channel. However, it is not possible for one agent to query another and retrieve these formulae, because a variable is only capable of representing a term, not a well-formed formula [36].

Enumeration mappings

may be represented as a set of rules, one for each value in the enumeration. For example, the following formulae state that if the value of a channel "Rl1" is 0.4, then the circuit breaker "X110" is closed. If the value of the "Rl1" channel is 0, then the circuit breaker "X110" is open.

$$Value(Rl1, 0.4) \Rightarrow Status(X110, Closed)$$
$$Value(Rl1, 0) \Rightarrow Status(X110, Open)$$

Many to one mappings may be represented by more complex mathematical expressions, for example, the following expressions state that if the value of "Rl1" is more than or equal to 0.2, then "X110" is closed, otherwise, X110 is open.

$$\forall x \, Value(Rl1, x) \wedge x \geq 0.2 \Rightarrow Status(X110, Closed)$$
$$\forall x \, Value(Rl1, x) \wedge x < 0.2 \Rightarrow Status(X110, Open)$$

Complex actions

may be represented as a sequence of actions (*ActionExpression* in the FIPA SL specification). For example, consider changing the tap position of a transformer on the substation simulator as described above. This is a four-step process:

1. Write 1 to the "tap up" channel of the correct data acquisition device.
2. When the tap change starts, the "tap change in progress" channel's value will change to 1.
3. When the tap change is completed, the "tap change in progress" channel's value will revert to 0.
4. The agent should then write a 0 to the "tap up" channel.

This process can be represented in logic by the following sequence of actions, assuming that the "write" action writes a value to a channel, and the "wait" action waits for a particular channel to have a particular value. The ";" operator represents the fact that one action is followed by another, as in FIPA SL and the logics described in [35, 118].

$$Action(Agent1, Write(tu1, 1));$$
$$Action(Agent1, Wait(tc1, 1));$$
$$Action(Agent1, Wait(tc1, 0));$$
$$Action(Agent1, Write(tu1, 0))$$

A BDI agent might wish to use this sequence of actions as a plan body. It would then be necessary to express the preconditions, postconditions, *etc.*, of this action sequence. Alternatively, we might define a primitive action, such as "tap-up" and then state that the complex action sequence "implements" the primitive action. For example:

$$ImplementedBy(TapUp(Sgt1), < action \; sequence >)$$

where<action sequence> is replaced by the sequence of actions specified above.

To express the preconditions of an action we could use (where <set of conditions> can be substituted by any legal first-order logical formula):

$$< set\,of\,conditions > \Rightarrow Feasible(< action\,sequence >)$$

This is legal because both arguments to *implies* may be well-formed formulae, the conditions are represented by the conjunction (*and*) of several well-formed formulae, and an action sequence in SL is a well-formed formula. For example, suppose we wish to state that it is possible to open a relay if it is not already open (the designation *Agent1* replaces the frame-based agent identifier that is used in FIPA SL):

$$\neg Status(Relay0, Open) \Rightarrow Feasible(Action(Agent1, Open(Relay0)))$$

The following rule is a general rule suitable for all relays, which states that for all a and r, if r is a relay and a manages r and the status of r is not "Open", then it is possible for a to carry out the action *Open(r)*. The *manages* predicate states that an agent "manages" or is responsible for a particular item of plant—in the system described this will be the plant agent corresponding to that item of plant.

$$\forall a, r\ Relay(r) \wedge Manages(a, r) \wedge Status(r, Open)$$
$$\Rightarrow\qquad Possible(Action(a, Open(r)))$$

Transmission of Mappings Between Agents

The mapping representations described above using *implies* are capable of being transmitted between agents (they are legal FIPA SL content expressions). However, it is not possible for an agent to query another to retrieve these mappings, because a variable in FIPA SL may only represent a term and not a well-formed formula. There are a number of possible ways to work around this limitation, including:

1. Use quotes to transmit the mapping expression as a string. It would then be possible to send a query, for example (all ?a (mapping-rule ?a)), and receive an answer such as (= (all ?a (mapping-rule ?a))) (set "(implies (value ds1 0) (status x10 open))")). The problem with this is that it is not possible to query mapping rules for an object involved (because the mapping rule is an opaque string), and so it is necessary for the mapping agent to maintain a database relating mapping rules to items of plant. For example, a database table might contain three columns: plant name, property name and mapping rule, in order that mapping rules could be searched.

2. Use a "mapping" structure to hold mapping information. (*e.g.* (mapping :channel TLV1 :scale-factor 3 :increment 2 :type "linear")). However, this only works for simple mappings, and it is also difficult to understand without prior knowledge of the mapping structure. Finally, the names of the mapping type (e.g. "linear") must be predefined, so it is not possible to represent arbitrary arithmetic expressions.
3. Use the "implements"/"implemented-by" relation as described above (only applicable to output mappings).

4.6 User Interface Agents

4.6.1 Description

As in other agent systems, the user interface agent in this system provides a link between the agent community and users. In this system, the interface with the user is via a HMI, different versions of which will be available for the substation and for the wide area network. The user agent must be able to translate knowledge from the representation used by the multi-agent system into that used by the human−machine interface. It must also be capable of carrying out tasks by using the directory services provided by the agent platform to locate appropriate agents to perform these tasks.

4.6.2 Agent Specification

- Knowledge
  - The agent should have a representation of the current state of the system, obtained at runtime from the other agents. It has no pre-programmed system knowledge.
  - If necessary, a set of mapping rules may be required to convert from the global system ontology to a representation used by the HMI.
- Capabilities
  - Generate configuration files for mobile data analysis and remote control agents, and launch agents.
  - Display information on a graphical user interface.
- Sensors
  - Input from a graphical user interface.
- Interaction Protocols
  - FIPA subscribe (initiator)
  - FIPA request (initiator)
  - FIPA query (initiator)
- Ontological knowledge
  - FIPA Meta Ontology, *plant* class from plant ontology. If mapping rules are used to convert information for display on the HMI, the power systems ontology will be used in these mapping rules.

4.7 Summary

This chapter has described the static agents used in the architecture presented in Chapter 3, along with an analysis of some of the design decisions involved. The static agents used in the system are database agents, document agents, ontology agents, device agents, plant agents, and user interface agents. All agents are based around a common agent architecture and consist of sensors, effectors (or abilities), and a knowledge base. Agents participate in one or more of the FIPA standard protocols as described previously.

The database agents and document agents are similar in that they both provide access to data sources. However, the methods used to query these agents differ in that a database is a machine-understandable data source which can be translated into first-order logic, while it is not possible to do this with a document. This affects the implementation of these agents, as the document agent must include an information retrieval algorithm for document ranking, while the database agent has the ability to translate between FIPA SL and SQL to query the database.

The plant agents, device agents and node agents make up the data acquisition and control system. The plant agents are capable of controlling and monitoring a single item of plant. Using the device agents, it is possible for a plant agent to do this without any knowledge of the control and monitoring hardware that this item of plant is connected to. Meanwhile, a device agent is responsible solely for a monitoring or control device, and has no knowledge of the plant. This separation of functions improves the modularity and flexibility of the system.

Design decisions involved in agent implementation include the representation of mapping rules used in the plant agents to convert from channel/value data to FIPA SL expressions. It was decided that these rules should be capable of representation in FIPA SL, in order that they could be stored in a database and transmitted to the plant agents when these agents were started, allowing dynamic agent configuration. However, it was not possible to query rules directly using first-order logic, and so a "mapping" frame was introduced in which the plant item to which the rule relates was specified, and the rule was provided as a quoted string. This allowed the plant agent to query mappings from the mapping database agent while maintaining compliance of the content of the query and reply with the FIPA specifications.

This chapter has considered the static agents present in the architecture. The next chapter will look at the use of mobile agents, which make up the remainder of the agents used. The chapter will examine whether these agents have the potential to improve the performance of some of the information management and control tasks that an automation system must perform.

5

Applications of Mobile Agents

The previous chapters have described the design of the architecture, and the static agents that are used to perform the various tasks of a power system automation system. However, the performance of such a system may be affected by the slow network links between the WAN and substations, and from one substation to another. For example, many of the current data links to National Grid Company substations operate at 64Kbit/s or 128 Kbit/s[1]. From previous research, such as [54, 56, 137], it can be seen that mobile agents are capable of providing a significant performance increase in applications consisting of multiple interactions or large data transfer over a low bandwidth or high latency network. This chapter therefore evaluates the tasks of data analysis and remote control, to determine whether their performance can be improved by the use of mobile agents [138].

5.1 Mobile Agent Performance

The model of mobile agent performance used in this chapter is based on that of Straßer and Schwehm [56]. When considering the performance of a mobile agent application, a number of factors must be taken into account:

- The amount of time to transfer a mobile agent a from a source host src to a destination host $dest$. We represent this by $T_{\text{transfer}}(a, src, dest)$. This time will depend on the speed and loading of the source and destination computers, the size of the mobile agent, the bandwidth and latency between the source and destination, and the mobile agent platform in use (different platforms employ different protocols for mobile agent transfer, some of which are more efficient than others).
- The amount of time to send a message m (in a client−server or message-based interaction) from a source host src to a destination host $dest$. We

---

[1] This information is based on discussions with NGC engineers and visits to substations.

represent this by $T_{\text{msg}}(m, src, dest)$. This will depend on the speed and loading of the source and destination computers, the message size and the bandwidth and latency between the source and destination. It will also be affected by characteristics of the protocol, and possibly the programming language or other tools, used.

- The amount of time taken to process a set of data or a request at a given server s. This is represented by $T_{\text{proc}}(s)$. It will depend on the speed and loading of the server and the particular processing operation or task being undertaken.

- The amount of time taken by a server program on server s to retrieve a requested data set from the database and prepare it for transmission to the client agent. This is represented by $T_{\text{ret}}(s)$, and depends only on the data set and the characteristics (performance and load) of s.

Time to Transfer a Message or Agent Across the Network

To obtain values for T_{msg} and T_{transfer}, the model defined by Straßer and Schwehm [56] may be used. In their paper, the time to perform a client–server interaction between two hosts L_1 and L_2 (involving a request and a reply) is defined to be equal to twice the latency δ between the two hosts, added to the total amount of data B_{RPC} transferred divided by the bandwidth τ between the two hosts, added to an overhead equal to 2μ multiplied by the amount of data transferred. The constant μ is determined by the software and computers in use. Therefore:

$$T_{\text{RPC}}(L_1, L_2, B_{\text{req}}, B_{\text{rep}}) = 2\delta(L_1, L_2) \tag{5.1}$$
$$+ \left(\tfrac{1}{\tau(L_1, L_2)} + 2\mu \right) B_{\text{RPC}}(L_1, L_2, B_{\text{req}}, B_{\text{rep}})$$

where

$$B_{\text{RPC}}(L_1, L_2, B_{\text{req}}, B_{\text{rep}}) = \begin{cases} 0 & \text{if } L_1 = L_2 \\ B_{\text{req}} + B_{\text{rep}} & \text{otherwise} \end{cases}$$

In this equation, $\tau(L_1, L_2)$ represents the bandwidth between L_1 and L_2, $\delta(L_1, L_2)$ represents the latency between L_1 and L_2 and μ represents the marshalling overhead, which depends only on the size of the object to be sent.

However, several assumptions made by this equation mean that it may not be suitable in all circumstances. The equation does not take account of the speed of the host computer, which affects the time to process a request and also the marshalling time. It can also be observed from results for Java object serialisation (which is the form of marshalling used by Java^{TM} RMI) in [139], that the time to marshal an object is not the same as the time to unmarshal that object.

In order to generalise the equation above, two functions are defined: $O_S(msg, src)$ to denote the sending overhead for message msg of size B_{msg}

from host src, and $O_R(msg, dest)$ to denote the receiving overhead for message msg at host $dest$. These are functions of the message (size and possibly content type) and of the particular host. The transmission time uses the simple $time = \frac{size}{bandwidth} + latency$ approximation, as can be observed by omitting the marshalling time from Equation (5.1).

$$T_{\mathrm{msg}}(m, src, dest) = \frac{B_{\mathrm{msg}}}{\tau(src, dest)} + \delta(src, dest) \qquad (5.2)$$
$$+ O_{\mathrm{S}}(msg, src) + O_{\mathrm{R}}(msg, dest)$$

The use of two overhead functions, one for sending and one for receiving messages, is proposed by Baldi and Picco [55]. However, they use "overhead" to mean the total time taken to send a message, while here the "overhead" represents only the cost of sending or receiving the message and does not include transmission time.

Time to Transfer a Mobile Agent

The actual time taken to transfer an agent over the network should be the same as for a message of the same size[2]. However, additional overheads are incurred by the time taken to register the agent on the destination platform and to restart its execution, *etc.* If these overheads are denoted by $O_{\mathrm{reg}}(agent, server)$ and $O_{\mathrm{dereg}}(agent, server)$, the resulting migration time equation is:

$$T_{\mathrm{transfer}}(a, src, dest) = O_{\mathrm{reg}}(a, src)$$
$$+ O_{\mathrm{dereg}}(a, dest) \qquad (5.3)$$
$$+ O_{\mathrm{S}}(a, src)$$
$$+ O_{\mathrm{R}}(a, dest)$$
$$+ \frac{B_a}{\tau(src, dest)} + \delta(src, dest)$$

In this equation, B_a represents the size of the agent, including code, data and state. As for the message, the main problem with this model is that different types of agent content may take differing amounts of time to transfer.

5.2 Mobile Agent for Data Analysis

In the architecture described in this thesis, data stored in a substation database is accessed using a multi-agent system [140]. This data is made

---

[2] In practice this may not be the case if, for example, additional messages must be sent between the agent platforms to co-ordinate the movement of the agent.

available via a database agent, which can be queried using FIPA ACL [95] messages to retrieve data. However, problems occur when it is necessary to analyse a large quantity of data, such as monitoring data covering a long period of time. This data can be extremely large, especially when converted into a text-based format, and the connections to substations are often slow (sometimes only dial-up links are available). Therefore, mobile agents are to be employed to analyse the data in the substation and remove the need to transmit it across the network. A subset of the generic architecture described in this thesis, consisting of two database agents per substation, an ontology agent, a user agent and the mobile agent, is used. This is shown in Figure 5.1. The monitoring database contains event data, and the static database contains substation topology information.

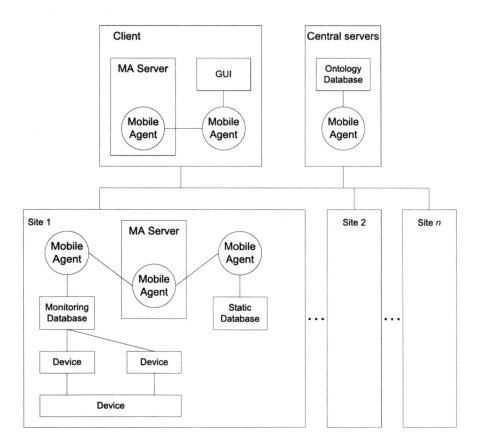

Fig. 5.1. Subset of generic architecture used for mobile agent-based data analysis

The basic outline of this application, as implemented in the prototype described in Chapter 6, is as follows:

1. The mobile agent is launched by the user agent and provided with a configuration file, consisting of a series of *data sets* that must be retrieved, a set of *analysis functions* that the agent must perform on those data sets, and a *report template*, specifying how the results produced by the analysis functions will be displayed in the generated report. Each data set consists of an object (*e.g.* a transformer), a property (*e.g.* LV current) and a time range (start time and end time). The process used by the user agent to generate the configuration file is described in Chapter 6.

2. The mobile agent gathers the specified data sets, and performs the required analysis, generating its report. For the multi-hop case, the implemented agent uses a basic planning algorithm to generate its route and the order in which its tasks will be performed.

3. The report is displayed to the user as an HTML page.

It is possible to create both multi-hop and single-hop data analysis agents. However, the single-hop agent is much simpler, as the agent only has to travel to a single location and perform a single analysis, and does not have to perform planning or route optimisation.

In the multi-hop case, the agent must analyse monitoring data from a number of substations. Because the data links between substations can be slow, it is desirable that the agent should transfer as little data as possible. Therefore, unless data from two substations must be used as a combined data set, the agent should perform all analysis on a substation's data, and retain only the result, before moving on to the next substation. It is desirable that the agent would be able to generate an optimal (or near-optimal) route through the substations to ensure that the tasks are completed in the shortest amount of time. Variations on this problem have been investigated in a number of papers, including an algorithm which attempts to optimise the number of agents used as well as the total time taken [141]. Moizumi analysed a similar problem, the "travelling agent problem (TAP)" in which an agent must perform a task involving a number of servers [142]. However, in the TAP, the task may be completed without visiting all servers, whereas in the application described here it is necessary for all relevant servers to be visited. Further work should investigate whether these algorithms can be applied to the agent described here.

5.2.1 Agent Algorithms and Implementation

Simple, Single-hop Analysis Agent

The single-hop data analysis agent can be implemented using a relatively simple algorithm, such as Algorithm 1. The agent travels to the location of the server providing the data set, retrieves the data, performs its analysis and sends the report to the user. The names of items of plant and properties which can be analysed can be obtained by querying the static and ontology databases respectively.

Some of the procedures used in this algorithm are dependent on the agent platform or on a particular implementation and are therefore not described here in detail. The **send** procedure delivers a message to a given agent. The **extract-results** procedure extracts a table of results from the reply to a query. The **location-of** procedure returns the agent platform, or container, on which a particular agent is located. The **kill-self** procedure terminates the agent's execution, freeing system resources. The **generate-report** procedure takes a set of results and a template, and generates a report.

Algorithm 1 Single-hop analysis agent

```
    Inputs:
        s: server agent
        e: expression defining data set
        template: report template
        F: set of analysis functions
    begin
        move(location-of(s))
        results := collect-data(e,s)
        report := generate-report
                    (results, template, F)
        display-report(report)
        kill-self()
    end

    procedure collect-data(e,s)
    begin
        message := create-query-message(e)
        send(message,s)
        reply := wait-for-reply(message)
        results := extract-results(reply)
        return results
    end
```

Multi-hop Agent Without Optimisation

The multi-hop agent performs a considerably more complex task than the single-hop agent. As well as retrieving and analysing data, it must be able to generate a plan (or sequence of actions) that ensures that all of the required tasks are completed and attempts to minimise the execution time. For example, if an agent must perform two tasks in substation 1 and one in substation

2, it is normally desirable (assuming that these are independent tasks) to perform either both tasks in substation 1 followed by the single task in substation 2, or the task in substation 2 followed by the tasks in substation 1. Compared to performing one of the tasks in substation 1, followed by the task in substation 2 and then the second task in substation 1, the number of "hops" that the mobile agent must make across the network is less. This problem has been analysed by Xie in [143], who proposed a scheduling algorithm for mobile agent planning based on Distributed Acyclic Graph (DAG) scheduling. However, this algorithm relies on the use of multiple mobile agents, and therefore is not used in the agent described here (a single agent).

In the case of the data analysis agent, there are four types of task which it may carry out: *retrieval*, which involves gathering a single data set from a server, *analysis*, which involves the use of a data set to generate either another data set or a result, *report generation*, which involves the generation of the report for the user, and *display*, which displays the generated report on the screen.

When the multi-hop user agent reads the configuration file, it first generates a task list from the data sets and analysis results defined in the configuration file. This task list will always contain one report generation task and one display task. For each data set a retrieval task is generated, and for each analysis an analysis task is generated. Following Xie's DAG representation, for each task a set of *predecessor* tasks (those which must be completed in order for the task to be performed) and a set of *successor* tasks (those which may only be started once the task is complete) are defined. In the mobile analysis agent, retrieval tasks have no predecessors. An analysis task has as its predecessors all those tasks (either retrieval or analysis tasks) providing input to the analysis. The report generation task can only be performed once all analysis tasks are complete, and the display task can only be performed once the report has been generated.

Once a plan has been generated, the mobile agent will then execute each step of the plan in order. The algorithm of the multi-hop mobile agent is given as Algorithm 2. Each task is assigned a location by the agent, and the *location_of* procedure retrieves the location of a specified task. In the basic agent described here, the location of a retrieval task is always the agent container on which the server agent for that retrieval is located (the agent will always move). The agent will carry out an analysis or report generation task at its current location, and will return home before displaying the report. The basic planning algorithm used in the prototype implementation (not shown) "clusters" tasks so that as soon as a data retrieval task is complete, all possible successor tasks (normally analysis tasks) will be performed. However, it does not currently cluster data retrieval tasks by substation−this optimisation will be implemented later.

Algorithm 2 Multi-hop analysis agent

```
Inputs:
    f: configuration file
begin
    config := read_configuration_file(f)
    tasks := generate_task_list(config)
    plan := generate_plan(tasks)

    while not complete(plan) do
        t := first_incomplete_task(plan)
        move(location_of(t))
        execute(t)
    wend
end
```

5.2.2 Benchmarks

A benchmark evaluation of the single-hop data analysis agent was performed. A data analysis agent was created, which retrieved a data table consisting of two columns (value and time) from a database and performed the following analyses:

1. Calculate the mean value of one column of data (single iteration through data retrieved).
2. Calculate the maximum value of one column of data (single iteration through data retrieved).
3. Plot a graph (two iterations through data retrieved).

Using a configuration parameter of the agent, it could be set to operate either as a static or mobile agent. This enabled a comparison to be performed. A second comparison was also performed between the use of wrapper agents, located at the server, to retrieve data, and an implementation which was granted direct access to the database, and used Java Database Connectivity (JDBC) for database access. In the first case, the database agent used the ACL message from the mobile agent to query the database, returning the results as an ACL message with FIPA Semantic Language (SL) content. In the second, to maintain the independence of the mobile agent from the data source, the agent sent an ACL request to the wrapper agent, which provided it with database connection details and the SQL query that could be used to retrieve the data. (For this experiment, the query and connection information were hard-coded into the database agent's configuration, as a full translation procedure to generate SQL, rather than directly query the database, has not yet been implemented.)

The client and server computers were connected to separate Ethernet switches and a computer running the Dummynet program [144] was used as a bridge to control the bandwidth and latency between client and server, as shown in Figure 5.2. A number of different settings of bandwidth and latency were used. The time taken for the agent to complete its analysis and display the results to the user was measured. This was repeated six times for each bandwidth/latency combination, three times in the mobile agent case and three times in the client−server case. While the use of Dummynet means that the network delay is deterministic, it is considered that this is appropriate for a power company network in which the wide area network consists of private wire circuits, which do not suffer from the random delays due to high traffic as exhibited by the Internet.

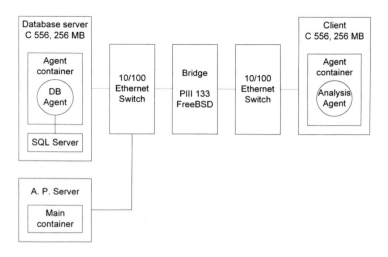

Fig. 5.2. Experimental setup

Based on the performance model discussed in Section 5.2, it is expected that the mobile agent-based data retrieval agent will perform significantly better than the client−server agent. This performance advantage should increase with decreasing bandwidth and with increasing latency.

Results

Rather than taking the mean of the three measurements, it was decided to use the first measurement for each set of latency/bandwidth values, as this includes the cost (in the mobile agent case) of transferring the agent's class files, which is not included in later measurements as these files are already transferred. Table 5.1 shows the fixed parameters for the experiment (object sizes were obtained by serialising the object into a byte array and then recording the length of the array).

Table 5.1. Fixed parameters for data retrieval experiment

Parameter	Value
Number of rows of data retrieved	18,915
Number of columns of data retrieved	2
Object size of wrapper-based mobile agent (outgoing)	7963 bytes
Object size of wrapper-based mobile agent (return)	8183 bytes
Report size of wrapper-based mobile agent (return)	4268 bytes
Size of reply message from wrapper (containing data)	1053525 bytes
Size of request message (wrapper-based case)	2249 bytes
Total size of wrapper-based agent class files	42950 bytes

Tables 5.2, 5.3 and 5.4 give the first run times in seconds for each of the six agents, for bandwidth from 100 Kbit/s to 100 Mbit/s, and one-way latency between 0 and 50 milliseconds. All times are to 3 significant figures.

Table 5.2. Time taken (seconds) by mobile direct access agents (times to 3 sf)

		1-way latency (ms)				1-way latency (ms)	
		0	25	50		0	25	50
τ(bit/s)	100 M	109	105	107	100 M	33.4	35.9	38.1
	10 M	102	104	104	10 M	33.1	36.1	38.2
	1 M	103	105	108	1 M	34.1	36.2	38.6
	100 K	110	112	114	100 K	41.8	44.3	46.0
		No caching				Caching	

Figures 5.3 and 5.4 show the results obtained by varying the one-way latency of the connection from 0 to 50 milliseconds, while leaving the bandwidth unchanged at 10 Mbit/s. Figure 5.4 shows in more detail the results from the four faster scenarios, which are obscured on Figure 5.3 by the long times taken by the static agents with direct database access.

Figure 5.5 shows the effect of bandwidth on the experimental database analysis, using all of the different agents and with no added latency.

Table 5.3. Time taken (seconds) by static direct access agents (times to 3 sf)

		1-way latency (ms)				1-way latency (ms)	
		0	25	50		0	25	50
$\tau(\mathrm{bit/s})$	100 M	96.7	3802	7590	100 M	37.6	958	1910
	10 M	103	3820	7610	10 M	32.6	961	1910
	1 M	158	3940	7730	1 M	46.0	992	1940
	100 K	1390	5180	8970	100 K	354	1300	2250
		No caching				Caching		

Table 5.4. Time taken (seconds) by wrapper-based agents

		1-way latency (ms)				1-way latency (ms)	
		0	25	50		0	25	50
$\tau(\mathrm{bit/s})$	100 M	113	113	117	100 M	110	116	123
	10 M	112	113	117	10 M	111	119	122
	1 M	115	114	116	1 M	118	123	126
	100 K	119	120	123	100 K	202	203	208
		Mobile				Static		

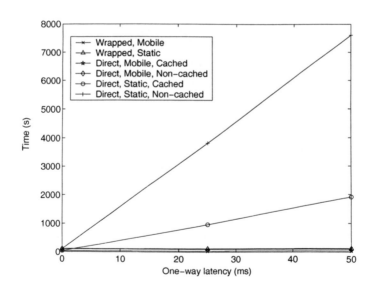

Fig. 5.3. All agents at 10 Mbit/s

Analysis

The experiment described here is relatively limited. However, there is a significant body of existing work in the field of mobile agent performance, some of

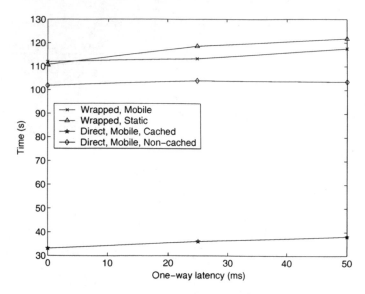

Fig. 5.4. Agents at 10 Mbit/s with static direct access agents removed

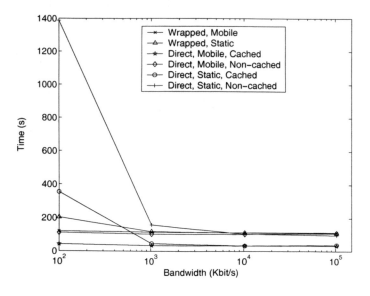

Fig. 5.5. Time taken to analyse database at various bandwidths, no added delay

which is discussed in Section 5.2.4. Therefore, the intention of the experiment described here is to confirm that the advantages found by existing work are applicable in the context of the power system automation architecture developed in this thesis, and to demonstrate that via a specific application used in the prototype. The work here does not attempt to develop a complete model of mobile agent performance.

Effect of Latency

A number of points can be observed from the data in Figure 5.3, which shows how the performance of the agents is affected by changes in latency:

- When the latency was zero (at the y axis of Figure 5.3), the time taken by both wrapper-based agents (static agent and mobile agent) was approximately the same.
- Of the direct access agents, the mobile and static cached agents both had similar performance at zero latency, as did the mobile and static non-cached agents.
- The direct access static agents are affected much more by latency than any of the other agents.
- Neither wrapper-based agent is affected greatly by latency.
- The time taken by the direct access, static, cached agent is between three and four times the time taken by the direct, static, uncached agent. It is likely that this is because the data was iterated four times. Therefore, four times as much data was transferred in the uncached case. However, some of the time taken was due to processing rather than data transfer, so the ratio is less than four.

This difference in performance between mobile and static agents at zero latency, if not an experimental error, might represent the setup and movement overheads of the mobile agent, or the fact that having the agent on the same server as the database increases the server load.

Effect of Bandwidth

Figure 5.5 shows that the time taken to perform the task increases as the bandwidth decreases in all cases, and that this effect is much greater for the static agents (due to the greater amount of data transferred).

5.2.3 Discussion

The results show that the performance of the client–server data analysis application is highly dependent on the bandwidth of the network, while that of the mobile agent is relatively unaffected. Therefore, by using a mobile agent it is possible to greatly improve the benchmark performance in low bandwidth cases. These results agree with the predictions made by the model of this application in Section 5.2. Equation 5.3 states that the difference between the time taken by a client–server program to perform the task and the time taken by a mobile agent is equal to the difference between the amount of time taken for the client–server program to send and receive the query and reply messages and the amount of time to transfer the agent. Because the reply message is much larger than the agent, as the bandwidth increases the time taken by the client−server program should increase more quickly than the time taken by the mobile agent program. This agrees with the results of the benchmark. However, because of the significant overheads (platform overheads and data retrieval time) involved in the benchmark, the actual time taken does not match that predicted by the equations, which do not include the data retrieval time, as discussed previously.

When using an ACL-based data analysis agent, because only a small number of messages are sent, performance is not affected as much by high latencies as by low bandwidth, at least for realistic latencies (a typical "ping" time when using a dial-up modem is around 200 milliseconds, which gives 100 milliseconds one-way latency). As only a small number of messages (two in the optimal case) are sent and received, latency should only slightly affects the agent performance. The fact that altering the latency produces a greater change in time taken than this suggests that a greater number of messages are sent across the network when transmitting a mobile agent or an ACL message.

When using direct database access, the performance of the client−server agent was heavily affected by latency. This might indicate that the agent had to send a request for each row of data. This lack of optimisation of the client−server program may exaggerate the performance advantage of using a mobile agent in this application. By retrieving more data in each call, it might be possible to improve the performance of the direct access static agent to more closely match that of the wrapper-based static agent. However, this would mean that it would still be slower than the mobile agent in many cases.

It would be possible to equal the performance of the mobile agent for data analysis in a client−server system by building a server containing all of the analysis functions, and calling these functions remotely. However, as discussed in [57], this would greatly reduce the flexibility of the system, and make it more difficult to add new analysis functions as this would require the server program to be modified and restarted.

If the task is modified, this may alter the results of this benchmark. For example, changing the size of data retrieved will affect both the mobile agent and client−server systems. However, following the results of Johansen [145], mobile agents would be expected to perform better in relation to client−server systems as the data size increased, and vice versa. Changing the number of iterations required through the retrieved data would particularly affect the non-caching agents, as they must retrieve all of the data on each iteration. The effect on the caching agents would be limited.

5.2.4 Related Work

Johansen [145] demonstrated the performance advantages of mobile agents for the analysis of large data sets, using a mobile agent to retrieve weather satellite data. Johansen provides a comparison of the performance of client−server and mobile code for different data sizes, which shows a number of performance improvements from using mobile agents. Johansen's work provided some of the motivation for the work described here, as it showed that it was possible to improve the performance of a data analysis application using mobile agents. The work in this thesis extends Johansen's work by applying mobile agent data analysis in the context of the specific automation systems architecture developed in Chapters 3 and 4, and by providing benchmarks at different bandwidths and latencies.

Tsukui *et al.* [146] used mobile agents to retrieve data from power system protection devices. As in our work, they utilised a mobile agent for analysis. In their application it was used to analyse fault records from devices. They also used mobile agents to gather status information from devices and to alter device settings. However, their work does not include a detailed analysis of application performance, or describe the agent implementation in detail.

Gray *et al.* [54] describe two applications of mobile agents, one of which is an information retrieval application. They test the scalability of a mobile agent system for performing document retrieval and filtering as the number of clients increases, and find that client—server solutions perform poorly with large numbers of clients because the network becomes overloaded. In contrast, mobile agent solutions may perform poorly if a server machine becomes overloaded. Therefore, they conclude that if the bandwidth is high it is usually preferable to use a client—server system, and when the bandwidth is low or the number of clients is high it is usually preferable to use a mobile agent system. These conclusions are similar to the ones obtained from the experiment described here. However, the experiment that they performed used multiple mobile agents and was for document retrieval, whereas the experiment described here is for a single mobile agent performing database analysis. Therefore, the additional results provided here are useful to confirm that the conclusions drawn by Gray *et al.* are still valid for this application.

5.2.5 Conclusions

This section has described the use of mobile agents for performing data analysis tasks within the architecture described in this thesis. A performance model has been derived to allow the quantitative analysis of this mobile agent application. Performance benchmarks have shown that where the bandwidth of the wide area network is low or its latency is high, the use of mobile agents can provide significant performance improvements compared to a client—server or static agent implementation. Results have also demonstrated that it is possible to improve the performance of the mobile agent by providing direct, rather than wrapper-based, access to a database and by caching the data retrieved. However, these results may relate only to the specific wrapper-based implementation used here. It may be possible to improve the efficiency of the wrapper-based agent using an alternative implementation method, although it is likely that, because the wrapper-based implementation would still involve packaging the retrieved data as ACL and then parsing the ACL data, the direct access agent would still have a performance advantage.

When building the system it was desired to adhere as closely as possible to the FIPA standards for multi-agent systems. This motivated the choice of the JADE platform for system implementation, as at the time it was the only publically available FIPA platform to support mobile agents. However, JADE's focus is not on mobile agents, and so it lacks some of the mobile agent related features of other systems (e.g. strong mobility, support for multiple versions of a Java class in the same container, inter-platform mobility). For the benchmark, this was irrelevant as intra-platform mobility could be used to simulate inter-platform mobility, and the visioning problem was not encountered as only one version of the benchmark agent was used at a time. For a full deployment of the system it might be necessary to extend the platform or to re-examine the available alternatives.

Further work on this system should involve the full evaluation of multi-hop data analysis agents. In addition to this, it is also intended to add document retrieval functionality.

5.3 Mobile Agent for Remote Control of Power Systems

Many substations in an electricity network are unmanned, and must therefore be controlled remotely. Currently this is normally done from a control centre, using dedicated network links to the substation. Here we investigate the possibility of using mobile agents as a control mechanism to allow users to remotely control the substation plant over a standard IP network.

For example, suppose that a substation contains a transformer, with a circuit breaker and earth breaker connected to the high voltage windings, and a circuit breaker and earth breaker connected to the low voltage windings. This partial substation layout is shown in Figure 5.6. The substation provides both a mobile agent server and a client−server interface which allow commands to be sent to substation devices. A user at a remote site intends that the following actions should be performed in the substation, as illustrated by the numbers on Figure 5.6:

- Open two circuit breakers to isolate the transformer (1 and 2)
- Measure the voltage across the transformer to ensure that it has been isolated (3)
- If the transformer is isolated, operate two earth switches (one on either side of the transformer) to earth it (4 and 5)
- When all of these operations are completed, display a message on the user's machine.

This could be achieved in one of two ways. First, it would be possible to use the client−server interface to execute all of the commands remotely. Suppose

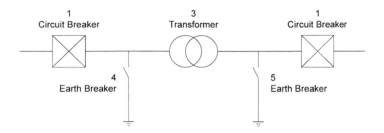

Fig. 5.6. Example substation

that the protocol is simple, and for each command only a single command message and a reply or acknowledgment message are required. In this case, 10 messages across the wide area network would be necessary (two for each of the two circuit breakers, two for the voltage measurement and two for each of the two earth switches). Alternatively, a mobile agent could travel to the substation and execute the same commands locally over the substation local area network. The 10 messages required in the client–server case would then be sent and received over the LAN. Only two messages would need to be sent across the WAN: the mobile agent and the final message informing the user that the sequence of actions was completed. Depending on the characteristics of the network involved, this might mean that the operation would complete more quickly if the mobile agent was used.

The major anticipated drawbacks to implementing this application are in the areas of security and reliability. Much has been written about the security concerns of mobile agents (*e.g.* [61]). However, these security problems mostly apply to open systems, in which it is possible for anyone to send a mobile agent to a server, and the servers are operated by different authorities. In a power system, which is relatively closed, these problems could, at least to a certain extent, be avoided. Also, the advantages of mobile agents for remote control might apply equally to less critical applications such as remote configuration management, in which, instead of instructing devices to operate substation plant, the mobile agent was used to alter device configurations. The performance model for this application is the same as that for remote control, as the agent is still performing a sequence of interactions with a device or devices, except that the interactions result in changes to the device configuration rather than changes to the status of the substation plant.

There is another problem when an agent is allowed to interact directly with IEDs. If multiple mobile agents are permitted to be present in the substation simultaneously, particularly when they are controlled by different users, it is possible that the commands transmitted by these agents might be inconsistent with one another. Also, it is not possible simply to "lock" individual items of plant being controlled, as commands executed on one part of the substation can affect other parts of the substation. For example, suppose a substation has two transformers. At the same time, user A sends a command to isolate transformer 1, and user B sends a command to isolate transformer 2. The result of this is that both transformers are isolated, and the substation then provides no output current.

It is proposed to solve this problem by using the plant agents as an intermediate layer between mobile agents and IEDs. When a mobile agent wishes to perform an action, it must request that the appropriate plant agent perform this action. By negotiation with other plant agents, that agent must then determine whether or not that action is possible, given any constraints which have been placed on the actions of the plant agents by other mobile agents or by human controllers. One possibility would be to adopt the Joint Intentions methodology, developed by Jennings for the ARCHON^{TM} industrial control multi-agent system [147]. Alternatives might include the use of a single, static, "supervisor" agent responsible for co-ordinating the actions of all agents in the substation, or the use of a constraint programming [148, 149] method.

5.3.1 Agent Algorithms and Implementation

For the experiment described in this chapter (Section 5.3.2), a simple (non-intelligent) mobile agent based on Algorithm 3 was implemented. The agent is given the address of a server agent (located at the closest agent server to the device), the URL of a relay device and a series of actions (either open or close) to be performed. When the agent starts, it will move to the location of the server agent and carry out the sequence of interactions with the device. The agent is capable of interacting with one device only. The preliminary benchmark used a similar algorithm, but without the server agent (the address of the destination agent platform was used instead).

Algorithm 4 shows a multi-hop agent, which acts through other agents, and uses the DF to determine which agents are capable of carrying out a particular action in its specified action sequence. The algorithm used in this case is as follows (in which the *location-of* procedure should return the substation in which either the item of plant affected by an action or a particular server agent is located):

The agent is given an ordered sequence of actions, which are represented by FIPA SL action expressions, giving the agent itself as the actor. For each of these actions, the agent first determines whether or not it is capable of performing the action itself. If it is, it will perform the action by directly interacting with the relevant IED. If not, the agent will attempt to perform the action using the capabilities of other agents. It first searches the DF, to find a list of agents (the variable *Servers* in Algorithm 4) capable of performing the relevant action. It will then request the action from each of these agents in turn, stopping when the action is complete. In a realistic implementation, it would also be necessary to provide the agent with the ability to handle

---

Algorithm 3 Mobile remote control agent

```
Inputs:
    s: name of server agent
    A: set of actions to perform
    d: device
    h: home location
begin
    l = locate(s)
    move(l)
    for each a in A do
        send-request(a, d)
        wait-reply(a, d)
    next a
    move(h)
end
```

Algorithm 4 Improved remote control agent

```
Inputs:
    A: sequence of actions to perform
    h: home location
begin
    for each a in A do
        if (capable-of(a)) then
            move(location-of(a))
            perform(a)
        else
            Servers = find-servers(a)
            for each s in Servers do
                move(location-of(s))
                request(s, a)
                if(success) then break
            next s
            if not complete(a) then fail
        endif
    next a
end
```

---

exceptional cases such as the failure of one of its actions (denoted in Algorithm 4 by `fail`).

$$T_{cs} = \sum_{j=0}^{n} (T_{\mathrm{msg}}(req_j, c, D_j)) + (T_{\mathrm{msg}}(rep_j, c, D_j)) + T_{\mathrm{proc}_j} \qquad (5.4)$$

$$T_{\mathrm{MA}} = \sum_{j=0}^{n} (T_{\mathrm{msg}}(req_j, s, D_j)) + (T_{\mathrm{msg}}(rep_j, s, D_j)) + T_{\mathrm{proc}_j} \qquad (5.5)$$
$$+ T_{\mathrm{transfer}}(ma, c, s) + T_{\mathrm{msg}}(results, s, c)$$

The processing time (T_{proc}) is unknown. However, the processing time is the same for both the mobile agent and client−server methods, and so can be omitted from the comparative analysis. Also, the equation will be simplified further by setting $T_{\mathrm{msg}} = \frac{\mathrm{size}}{\mathrm{bandwidth}} +$ latency, and by supposing that all request messages have the same size B_{req} and all reply messages the same size B_{rep}. While this ignores marshalling overheads, accurate measurements of these are not available, and in any case, the marshalling overhead should be small compared to the actual message/mobile agent transfer time.

For the client−server method, the equation is simplified to:

$$T_{\mathrm{CS}} = \frac{n(B_{\mathrm{req}} + B_{\mathrm{rep}})}{\tau_{\mathrm{wan}}} + 2n\delta_{\mathrm{wan}}$$

For the mobile agent method, the simplified equation is:

$$T_{\mathrm{MA}} = \frac{n(B_{\mathrm{req}} + B_{\mathrm{rep}})}{\tau_{\mathrm{lan}}} + 2n\delta_{\mathrm{lan}}$$
$$+ \frac{2B_{\mathrm{MA}}}{\tau_{\mathrm{wan}}} + 2\delta_{\mathrm{wan}}$$

In these equations, τ_{wan} represents the bandwidth of the wide area network, δ_{wan} represents the latency of the wide area network, and τ_{lan} and δ_{lan} represent the bandwidth and latency of the local area network. Because in the experiment the mobile agent returns to the client to display its results, the amount of data transferred in the mobile agent case is twice the size of the agent, rather than the size of the agent added to the size of an acknowledgment message.

The "crossover point" at which using a mobile agent becomes more efficient than a client−server system then depends on the bandwidth τ, latency δ, number of interactions n and the size of the mobile agent and of the replies. The sizes of the request and reply messages are unknown. As an approximation, it is assumed that each message has size 100 bytes. The size of the mobile agent is 6353 bytes (object) + 10796 bytes (class files). For these calculations, it is assumed that the class files are already present, and the object size only is used.

The three graphs, Figures 5.8, 5.9 and 5.10, show the theoretical time taken against the number of interactions, for several different settings of WAN bandwidth and latency. It can be seen from Figures 5.8 and 5.9 that decreasing the bandwidth of the WAN has the following effects:

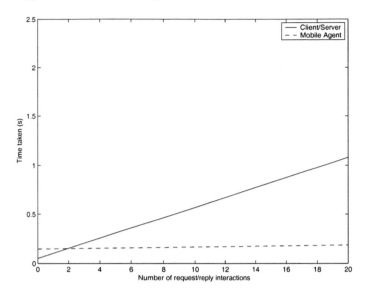

Fig. 5.8. Theoretical timing (excluding marshalling and operation time) for 1 Mbit/s, 25 ms WAN, 100 Mbit/s, 1 ms LAN

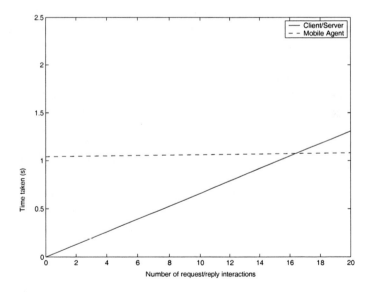

Fig. 5.9. Theoretical timing (excluding marshalling and operation time) for 100 kbit/s, 25 ms WAN, 100 Mbit/s, 1 ms LAN

- The gradient of the client/server line is increased.
- The intercept of the mobile agent line is increased.

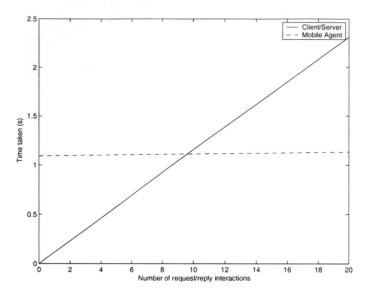

Fig. 5.10. Theoretical timing (excluding marshalling and operation time) for 100 kbit/s, 50 ms WAN, 100 Mbit/s, 1 ms LAN

Due to the relative magnitudes of these changes (the client/server messages are much smaller than the mobile agent) the overall effect is to significantly increase the number of interactions required before the mobile agent is the most efficient interaction method.

It can be observed from Figures 5.9 and 5.10 that increasing the latency of the WAN (by 25 milliseconds) has the following effects:

- The gradient of the client/server line is increased.
- There is little effect on the mobile agent line (actually, the intercept will be increased by 50 milliseconds, but this is not noticeable on the graph).

Therefore, the effect of increased latency is to decrease the number of interactions required before the mobile agent is the most efficient interaction method. However, the number of interactions will never fall below 1, as two messages (equivalent to a single interaction) must always be sent across the WAN when the mobile agent is used. Therefore, for a single request/reply interaction, the mobile agent will always be less efficient (as long as the combined size of the request and reply messages is smaller than that of the mobile agent), due to a larger *size/bandwidth* term.

Problems with the Estimated Crossover Point

The estimated "crossover point" will be optimistic (in favour of the mobile agent) because the mobile agent transfer overheads have been omitted. Also,

5.3.2 Experiment

In this experiment, an agent was used to remotely control a network-attached relay for read/write interactions. Because of the (proprietary) protocol used by the relay, each of these interactions resulted in a number of messages being exchanged between the agent and the relay. As in the data analysis experiment, the bandwidth and latency were varied using a bridge running the Dummynet program [144]. The system setup is shown in Figure 5.7.

The experiment was designed only to evaluate the use of mobile code in this application. Therefore the interactions were performed using a software library, rather than via static agents as described earlier in this book.

Some problems were encountered with the Java library that connected to the relay. In order to prevent these problems, a call to `Thread.sleep (500)` was added after each operation, causing the agent to sleep for 500 milliseconds. The agent also slept for 300 milliseconds after connecting to the relay. To obtain the final results, the total sleep time ($0.5n + 0.3$ seconds) was calculated and subtracted from the measured time. This might introduce a small error due to the inaccuracy of the `Thread.sleep` function. However, this should be reduced by the number of interactions used.

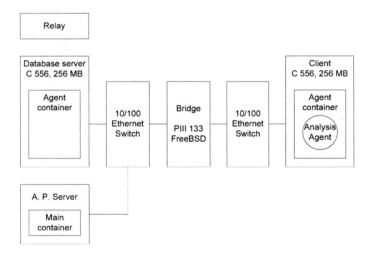

Fig. 5.7. Experimental setup for remote control experiment

Theoretical Results Sort Out

Using equations (5.4) and (5.5), it should be possible to predict the point at which the mobile agent method becomes more efficient than the client−server method.

the JADE platform is not as efficient as is theoretically possible, as a number of messages, rather than a single message, are sent in order to send or retrieve an agent. Therefore, it is expected that in the experiment the overall trends should be the same as the theoretical results described, but that the crossover point and time taken will occur at a slightly greater number of interactions. The time taken in all experiments will also be greater than the theoretical time due to the relay operation time and the time taken to connect to the relay.

Results

Table 5.5 shows the fixed parameters for the experiment. The latency between the mobile agent server and the relay and that between the client and relay at 0 latency were measured using a free "ping" implementation[3] and are the average of 10 pings. The size of the mobile agent is slightly inaccurate as the agent had to be modified in order to print out its size. However, this should only make a small difference to the object size. The class file sizes are those from the actual agent used in the experiments.

Table 5.5. Fixed experiment parameters

Object	Value
$\delta(s,d)$ (server$-$device latency)	1.13 ms
$\delta(c,s)$ when set $\delta(c,s)=0$ (client$-$server latency)	0.82 ms
$\delta(c,d)$ when set $\delta(c,s)=0$(client$-$device latency)	1.68 ms
Size of agent object	6353 bytes
Total size of agent .class files (3 files)	10796 bytes

An initial experiment (not described here) varied the number of actions between agent and relay from 0 to 5. However, it was found that with this number of interactions the client$-$server method always outperformed the mobile agent. Because the results of the first experiment did not show any point at which mobile agents became more efficient than the client$-$server method (the client$-$server method always completed the task more quickly than the mobile agent), a second experiment was performed. It was decided for this experiment to focus on the parameters of latency and number of interactions. Therefore, the bandwidth was fixed at 1 Mbit/s, the one-way latency was varied between 0 and 100 milliseconds and the number of interactions was varied between 0 and 100. The results of this experiment (as the mean time of four runs) are summarised in Tables 5.6 and 5.7. The sleep time between interactions and after connecting to the device has been subtracted from both

---

[3] www.cfos.de/ping/ping.htm

mobile agent and client server results, as this time was the same for both agents and should not be present in a full implementation.

Table 5.6. Time (seconds) taken by mobile remote control agent (mean time − sleep time)

		Number of interactions					
		0	20	40	60	80	100
	0	2.36	2.79	3.36	3.64	4.07	4.57
	25	3.87	4.22	4.75	5.23	5.63	6.12
Latency (ms)	50	5.32	5.71	6.32	6.83	7.31	7.61
	75	6.72	7.21	7.79	8.48	9.05	9.39
	100	8.22	8.92	9.36	10.02	10.64	11.00

Table 5.7. Time (seconds) taken by static remote control agent (mean time − sleep time)

		Number of interactions					
		0	20	40	60	80	100
	0	1.63	2.09	2.47	2.92	3.31	3.75
	25	1.98	3.30	4.78	6.07	7.50	9.00
Latency (ms)	50	2.53	4.84	7.31	9.64	12.05	14.47
	75	2.95	6.21	9.57	13.01	16.39	19.79
	100	3.30	7.53	11.88	16.29	20.69	25.12

Because of the implementation of the mobile agent, it was necessary for it to locate the server before moving. This was done by giving it the name of an agent on the server (representing a device agent) and then having the mobile agent ask the AMS for the location of this agent, which could then be used to request a move. This interaction used the FIPA request protocol, which involves three messages, a request message from the agent to the AMS, followed by an agree message and an inform message from the AMS to the agent.

In order to quantify the cost of this operation, another series of experiments were performed. The mobile agent was modified to print out both the total time taken, and the time taken excluding the time to locate the server. The time taken to locate the server is then the difference between these two times. The mean time to locate the server for four interactions, at different latencies, is shown in Table 5.8.

Table 5.8. Cost of locating server

Latency (ms)	0	25	50	75	100
Time to locate server	0.13	0.34	0.53	0.74	0.94

Analysis

Figure shows how the crossover point at which the time taken by a mobile agent became less than that taken for a static agent varied as the bandwidth and latency were altered. The area above the line is that for which a mobile agent provided superior performance.

It can be seen that for small numbers of interactions, or where the latency was low, the client–server, or static agent, method was superior. However, when the number of interactions and the latency were large, the time taken by the mobile agent was less than that for the static agent.

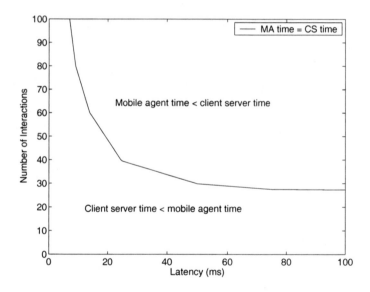

Fig. 5.11. Crossover point

Comparison to Theoretical Model

Using the model discussed earlier, the experimental results are now compared with theory, first for a scenario in which the latency is altered, and then for a scenario in which the number of interactions is altered. While it is not expected that the theoretical results will match the experimental results completely, due to the simplifications made in the formula and also due to the startup cost (starting the agent, loading a library and connecting to the relay) incurred in the experiment, it is expected that there should be a strong correlation between the two sets of results. In the theoretical results presented here, it is assumed that the latency of the LAN is 0 and the latency of the WAN is equal to the set latency. Using the actual values should not make a significant difference, as the latency of the LAN is actually less than 1 millisecond, giving an error for 100 interactions of 100 milliseconds or 0.1 seconds.

Effect of Altering Latency

By taking the client−server results for $n = 40$, the graph shown in Figure 5.12 is obtained. This shows that the theoretical and experimental results do not agree completely. First, there is a fixed cost in the experimental results (shown by the point at which $\delta = 0$). This consists of the time to load the library used to communicate with the relay, the processing time, or response time, of the relay, and the per-interaction overheads O_S and O_R. The difference in the gradients of the two lines (overhead varying with latency) is probably due to the time taken to exchange the messages required to establish a connection to the relay.

Using a linear approximation, the equation of the measured line is $t = 94\delta + 2.5$. The equation of the theoretical line is $t = 80\delta + 6.1 \times 10^{-5}$.[4]

Performing the same procedure for the mobile agent, and taking the mobile agent size to be 6353 bytes (its size excluding class files) the graph of Figure 5.13 is obtained.

This graph shows a significant difference between the predicted and experimental results for the mobile agent method. Even when allowing for processing time, mobile agent marshalling and registration overheads and the actual latency of the local area network by shifting the mobile agent line so that the values at $\delta = 0$ are identical, there is a marked difference in the gradient of the two lines. The model predicts that, because there should be only one message sent from client to server and one from server to client (to transfer the mobile agent), the latency of the wide area network should have little effect on the time taken (if the latency is 100 milliseconds, then this should add only 200 milliseconds to the total time). If the recipient of a mobile agent must send a message to acknowledge the mobile agent, the effect of latency should double, which would double the gradient of the line. However, this would still not be close to the experimental results. In the experiment, there is a difference of approximately 6 seconds between the time taken when $\delta = 0$ and the time taken when $\delta = 100$, when even allowing for a request and a reply between the mobile agent sender and recipient the difference in theory would be 400 milliseconds.

---

[4] Because $\frac{B_{req}+B_{rep}}{\tau(d,c)}$ is insignificant compared to the rest of the equation (the bandwidth is 1 Mbit/s), which approximately equals 128 Kbyte/s, changing the (estimated) values of B_{rep} and B_{req} makes little difference to the theoretical results.

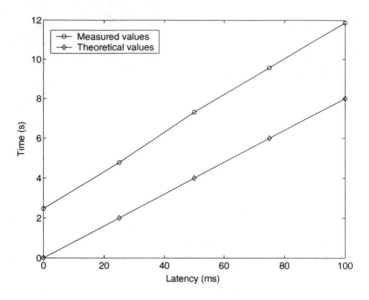

Fig. 5.12. Comparison of theoretical and actual results for static agent with $n = 40$

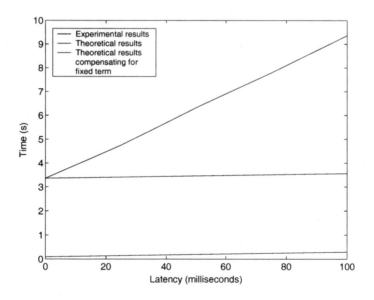

Fig. 5.13. Comparison of theoretical and actual results for mobile agent with $n = 40$

Part of this difference is undoubtedly due to the need to locate the server before the mobile agent moves. As shown by Table 5.8, the time to do this ranges from 0.13 seconds at 0 latency to 0.94 seconds at 100 milliseconds one-way latency. However, this is insufficient to account for the whole difference.

Another possible reason might be the inefficiency of the Java RMI protocol, used by JADE for message and mobile agent transfer. According to [150], it is possible for an interaction using RMI to require as many as six round-trip interactions for a single request and reply. There may also be other unknown overheads in mobile agent transfer, which result in message exchange.

Changing the Number of Interactions

Figure 5.14 shows the results for a set latency of 50 milliseconds, varying the number of interactions. As for when the number of interactions was fixed, there is a large fixed overhead due to the time taken to start the agent and connect to the relay. There is also a variable overhead which increases with the number of interactions. We hypothesise that this is due to the marshalling overheads and to the relay's operation time. For the measured results, a linear approximation is $t = 0.12n + 2.5$. For the theoretical results, a linear approximation is $t = 0.1n$. Therefore, for this case, the variable overhead is equivalent to $0.02n$, or 20 milliseconds/interaction, which is a realistic value for T_{proc}. However, as the value is so small it is also possible that it is affected by experimental error.

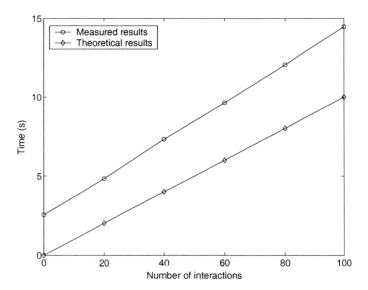

Fig. 5.14. Comparison of theoretical and actual results for static agent with one-way latency = 50 ms

For the mobile agent, the graph obtained is shown in Figure 5.15. The discrepancy between the fixed term in theory and in practice can be explained by the increased cost (compared to the theoretical value) of moving the mobile agent and to the cost of establishing a connection to the relay, as observed

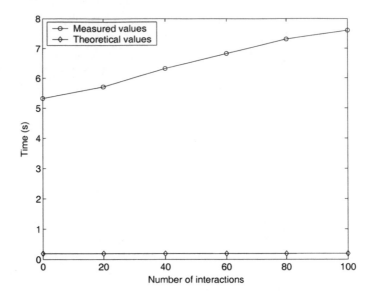

Fig. 5.15. Comparison of theoretical and actual results for mobile agent with one-way latency = 50 ms

in previous results. The difference in gradient can be attributed to the processing time T_{proc} and to the fact that the actual latency of the LAN was 1.13 milliseconds, and not 0, which would create an additional time cost of approximately 2.2 milliseconds per interaction. A linear approximation to the plot for the measured data is $t = 0.024\delta + 5.3$, whereas that for the theoretical data is $t = 1 \times 10^{-5}\delta + 0.19$. However, the gradient of the measured data plot is only slightly larger than the suggested value for $T_{\text{proc}} = 0.02$ seconds per interaction obtained from the plot for the static agent. This suggests that the main problem with the theoretical model for the mobile agent based method is that, as previously discussed, it significantly underestimates the time to move the mobile agent.

Discussion

One drawback of using a mobile agent for remote control is the fact that the actions to be performed must be known to the operator at the time the agent is launched. If an agent is only to perform a single action, its performance is no better than the client−server case. This means that the agent cannot be used, for example, for cases where the operator must select an action to perform based on the results of the previous action. However, in circumstances when a sequence of actions can be determined prior to launching the agent, such as the example given at the start of this section, the mobile agent does provide appreciable performance improvements.

In the experiment described here, the "crossover point" between mobile agent and static agent implementations came at a relatively high number of interactions (around 30−40 interactions for 25 ms<latency<100 ms). This is probably because the time taken by the mobile agent to locate the server, move and connect to the relay (shown by the results for $n = 0$) represents a high proportion of the total time taken in all of the interaction sequences. In contrast to this, the "startup time" for the static agent was much less, as it had only to connect to the relay. In order to make mobile agents more suitable for small numbers of interactions, effort should be made to reduce this overhead. This problem is similar to one highlighted in other work [57].

5.3.3 Conclusions and Related Work

The experimental results permit the conclusion to be drawn that mobile agents can improve the performance of a sequence of remote control operations over a high latency network. However, the exact sequence length and latency required before mobile agent performance is better than client−server performance depend on the particular implementations of the agent, agent platform and control devices. For example, in a previous experiment described in [151], only a very small number of interactions were required under high latency/low bandwidth conditions before the mobile agent's performance was superior to that of the client−server system. This was due to the particular conditions of that experiment.

When building the system, which has a multi-agent component as well as a mobile agent component, it was desired to adhere as closely as possible to the FIPA standards. This resulted in the choice of the JADE platform for system implementation. However, JADE's focus is not on mobile agents, and so it lacks some of the mobile agent related features of other systems (strong mobility, support for multiple versions of the same agent class in the same container, inter-platform mobility), and its agents are also more "heavyweight" (around 30 kilobytes for a basic serialised agent, excluding class files which, as part of the system, were already present). For a full deployment of the system it might be necessary to extend the platform or to re-examine the available alternatives.

Another problem encountered in constructing the system was that much of the industrial automation software and hardware that was to be integrated with the mobile or static agents is Windows based and communicates only via ActiveX® or via DLL libraries. Therefore considerable effort had to be undertaken, using the Java Native Interface (JNI), to allow agents to access these applications. Further work in this area would benefit from the development of simpler methods of accessing industrial automation devices using Java, or from the use of alternative technologies for the implementation of multi-agent systems (for example, [152] discusses the porting of a mobile agent system to Microsoft.NET®).

Tsukui *et al.*[146] have used mobile agents to retrieve data from power system protection devices. As in our work, they utilised a mobile agent for analysis. In their application it was used to analyse fault records from devices. They also used mobile agents to gather status information from devices and to alter device settings. This application is similar to the control application described here, as the alteration of a setting on a device can be considered, from the point of view of an agent, as a control operation (the agent sets the setting and receives a reply to confirm that it has been altered). However, their work does not include a detailed analysis of application performance, or describe the agent implementation in detail.

Harrison, Chess and Kershenbaum [58] suggested the use of mobile agents for remote control applications, and suggest that agents facilitate "remote real-time control when the network latency prevents real-time constraints being met by remote command sequences". However, they do not provide a performance model or any experimental results. The results of the experiment described here suggest that the statement made by Harrison *et al.* may be correct.

5.4 Summary

This chapter has examined the application of mobile agents within the architecture of Chapter 3, concentrating on the applications of data analysis and remote control. In the data analysis application, a mobile agent was used to travel to a database and perform some analysis on the data in that database, before returning a report to the user. An extension of this agent was capable of analysing multiple data sets, and was implemented in such a manner that new analysis functions could be programmed by the user. In the remote control task, an agent was used to carry out a sequence of actions in a substation, via the control devices. It was demonstrated that in the data analysis task, the use of a mobile agent provided superior performance to the use of static agents, provided that the amount of data transferred was large or the bandwidth of the network was low. In addition to this, it was demonstrated that when using a mobile agent, a performance improvement could be achieved by providing direct JDBC access to the database rather than by using a wrapper agent to perform translation. However, data source independence could still be achieved by allowing a wrapper agent to be used to generate the SQL query, which was then passed to the mobile agent. This technique is particularly appropriate due to the simple nature of the data sets used by the analysis agent (flat tables consisting of two columns), and may not work in more complex data retrieval tasks. For control tasks, it was demonstrated that the mobile agent had higher performance than a client−server control system when the number of interactions to be performed was large and the latency of the network was high. However, because control messages are typically small, reducing the bandwidth of the network, at least for smaller numbers of interactions, decreased the relative performance of the mobile agent as the agent size exceeded the total size of the control messages.

The next chapter presents a prototype implementation of the full architecture, including both mobile and static agents, and provides examples of its usage. Based on a real-time simulator provided by the National Grid Company, the prototype provides a realistic testbed for the concepts described in the book so far.

6

Multi-agent-based Substation Information Management System

6.1 Introduction

This chapter describes a distributed substation automation system prototype developed using the architecture and models described in this thesis. To provide a realistic test environment for the system without the difficulty involved in an implementation on an operational site, a real-time substation simulator provided by the NGT was used [153], as shown in Figure 6.1. This simulator consists of an industrial computer with a large number of analogue input and output channels, which are identical to the I/O facilities available to substation controllers in a current substation [154]. Because the simulator was designed to test substation control equipment, any system which operates correctly with the simulator should be capable of doing so in an actual substation environment.

The tasks of the prototype system are to gather data from the substation simulator via a data acquisition system, store this data in the National Grid Information Management Unit (IMU) database, and provide online display of data, historical data querying, data analysis and documentation management services to users via a human−machine interface.

The prototype is implemented using the JADE multi-agent systems toolkit [71]. The reasoning engines of the agents are implemented using a Prolog interpreter[1] with a Java interface. FIPA SL expressions (queries and requests) may be converted into Prolog expressions in a relatively simple manner, as both are based on first-order logic, and the main difference lies in the syntax (although FIPA SL has modal expressions and frames which are not implemented by Prolog).

---

[1] AMZI Prolog (http://www.amzi.com).

Fig. 6.1. National Grid Transco substation simulator

6.2 System Architecture and Agents

The information management system, shown in Figure 6.2, uses the generic architecture described elsewhere in this thesis. However, the generic data logging database has been replaced by the Information Management Unit (IMU), which performs the same functions but represents the actual system to be installed in National Grid Company substations. The IMU is based on a Microsoft SQL Server database, and has a Web service interface which allows other programs to use the Simple Object Access Protocol (SOAP), a World Wide Web Consortium standard, over Hypertext Transfer Protocol (HTTP) to insert data into the IMU or to perform queries. The component of the IMU used to acquire data is the Data Recording Service (DRS), and that used to query the IMU is known as the Query Service. There is also a second interface to the IMU via the Microsoft .NET®Removing protocol.

Figure 6.3 shows the architecture of the information management system, based around the IMU system. Data is gathered from the substation simulator by a data acquisition PC, and stored into the IMU. The IMU is managed by its own agent, the IMU agent, which is based on the database agent described in Section 4.1. The other agents shown on the diagram are as given on the architecture diagram of Figure 3.7 and described in Chapters 3 and 4. However, the mobile server (MS) agent has been omitted, as no mobile servers are available in the prototype implementation. Also, there are currently no task agents present in the prototype.

Because the IMU is only a database, it is not possible to control the system via the IMU. Also, it does not provide any means to automatically notify a subscriber when data in the database is updated. This means that it is difficult to support event-driven updating, for example, the FIPA subscribe protocol,

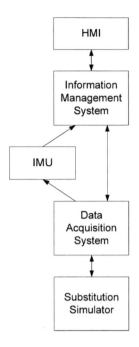

Fig. 6.2. Overall structure of system

using the IMU. Therefore, the agent-based data acquisition and control system shown in Figure 6.4, which uses the architecture described in Section 3.2, is used for control and event-based data updates.

The system contains only a single IED, the data acquisition PC system. Although this PC contains several I/O cards, it communicates with the multi-agent system over a TCP/IP connection, and it is simpler to treat the PC as a single device. Therefore only one device agent (represented in Figure 6.4 by "DAQ Agent") is present in the system. There are multiple plant agents (in a complete system one for each item of plant, however, in the current prototype only three have been implemented), co-located with the device agent and acquiring data by communicating with it. These agents are capable of passing information to the user interface agent, and of carrying out control commands sent to them.

6.2.1 Information Management System Agents

The information management system consists of database agents (as described in Section 4.1), an IMU agent for communication with the IMU, user interface agents and mobile agents. In addition, broker or task-oriented agents may be used to provide specific services. Further detail on the implementation of these

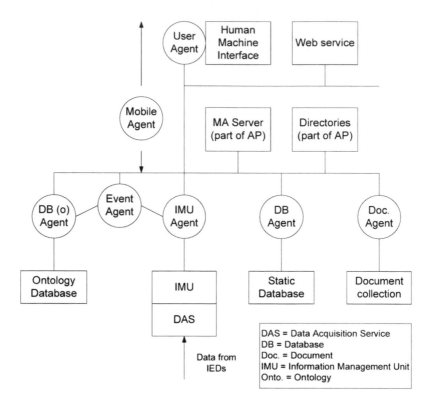

Fig. 6.3. Information management agent system architecture

agents is provided by Appendix B, and the format of the configuration files used is described in Appendix C.

Database Agents

There are three database agents present in the system. The *static database agent* manages a database containing static configuration information regarding the substation plant and the data acquisition system. The *mapping agent* (part of the data acquisition system) manages the mapping rules allowing agents to perform input data interpretation. The *ontology database agent* manages a database containing the system ontologies. The behaviour of these agents is as described in Section 4.1.

The IMU Agent

While the National Grid Company IMU is a database, it is not accessed via an SQL-based interface, but via a Web service. Therefore, an additional agent, the *IMU agent*, is required for this system. The reasoning engine of this agent

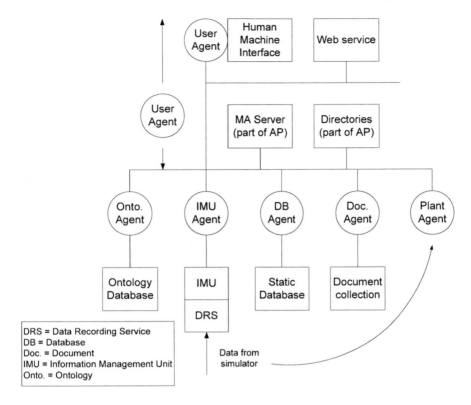

Fig. 6.4. Data acquisition agent system architecture

has been modified to add procedures which allow the Prolog interpreter to query the IMU via the Web service interface. However, to external agents, this agent will appear the same as it would if it were a database wrapper agent, as it still uses the FIPA Query protocol and the global ontology. The mapping rules of the IMU agent, written in Prolog, translate between the global system ontology and the schema used by the IMU's Web service interface.

The Alarm/Event Agent

The alarm/event agent is a *broker* agent that takes events and alarms generated by plant agents and forwards them to user agents as appropriate.

It operates by locating agents providing a subscription service and then advertising the subscriptions provided by those agents as if it provided them itself. When an agent subscribes for a particular event notification, the alarm and event agent establishes a subscription with the provider agent and forwards any event notifications received.

The use of a broker agent provides the possibility to add new features, for example, to generate alarm conditions by combining information from multiple sources. This functionality has not yet been implemented in the prototype and is a topic for further work.

User Interface Agent and HMI

Agents can be designed to provide a friendly interface for user accessing information resources and the capabilities of observing user's preference optionally to satisfy the user's specific requirements. They are commonly called *user agents*. The roles defined in this community are usually concerned with their distributions to implementing additional functional specifications of the system. There are also a number of different implementation methodologies for the user agents to provide varying degrees of intelligence. There are two types of user agents which provide individual services: one is the User Interface Agent (UIA) which provides a user interface to the system that allows its implementation details to be hidden from the user. For example, A UIA may be integrated into an HMI package or shown on a Web page. The UIA is implemented, in the substation information management system, in VisualBasic® (Microsoft Corporation, Redmond, WA) and provides a link between the HMI interface (implemented in LabVIEW™ (National Instruments, Austin, TX)) and the multi-agent system. It also provides its own graphical user interface for generation of mobile data analysis agents and database querying. The other is the personal assistant agent which will be discussed in the following section.

6.2.2 Personal Assistant Agents in Substation Information Systems

The personal assistant agent (PAA), $agent_{PAA}$, derived from the user agent role, provides a personal link between the agent community and users. In order that a PAA is able to provide a personalised service to the user, it must maintain a profile containing information about the user's interests and typical information requirements. This allows the PAA to select information that is most appropriate to the user. This version of $agent_{PAA}$ includes the following services:

- register with agent platform. Users are able to find out which agents are ready to provide services;
- subscribe information/services from other agents;
- switch on/off a relay by coordinating the ontology agent and device agents;
- ontology-driven information queries;
- query information directly from the IMU and generate reports;
- search documents for users;
- provide a thermal model, based on transformer real-time condition monitoring;
- web-based access to transformer thermal analysis results;

- download archive data;
- transformer evaluation.

In the substation information management system, the $agent_{PAA}$ is developed in C++ language to achieve a more user-friendly interface. The $agent_{PAA}$ implements the FIPA subscribe, request and query interaction protocols. More details of the usage of PAAs will be presented in Section 6.4.6.

6.3 System Ontology

As described in Section 3.5.3, the system's ontology consists of several components: the *automation ontology*, which provides a (partial) generic ontology for data acquisition and control systems, the *substation plant ontology*, which describes the different items of plant found in substations, and the *information management ontology*, which describes documents, other information resources and querying [155]. The instantiations of those ontologies used in the development of the prototype are described here. The ontologies are modelled using a UML class diagram notation, with operations representing actions that can be performed on an instance of a particular class.

Automation Ontology

The automation ontology describes industrial automation devices and systems. In the prototype system, the ontology used is that shown in Figure 3.5, which was used to design the data acquisition multi-agent system and is described in Section 3.3.2.

Plant Ontology

The plant ontology describes substation plant and its properties, along with the operations that may be performed on it. It is not intended to represent all of the properties available in a full substation automation system which would require a substantially larger ontology, but only to provide an example and a simple ontology for the prototype implementation. This ontology is based on the properties provided by the substation simulator, and is shown in Figure 6.5. Additional properties of the transformer object have been taken from the database schema of the transformer monitoring system described in [125]. However, in this schema, there are quantities representing both actual and predicted quantities, for example OIL_IN_ORG (actual oil in) and OIL_IN_OUT (predicted oil in temperature). For the purposes of the ontology, it is considered that the predicted temperature is not actually a separate property, but that it represents what some agent believes that the value of that property will be at a certain time. The ontology represents only actual properties of the transformer. However, this is not the only way to represent the transformer. An alternative would be to represent the transformer as a collection of components (*e.g.* tap changer, windings, tap change miniature circuit breaker), and then represent the properties of these components.

All properties of the disconnecter and circuit breaker are inherited from the parent class, "switchgear". This is because the difference between these types of plant is that it is not possible to open a disconnecter while it is live. This cannot be shown in the object model and should instead be encoded into the behaviour of the relevant plant agents.

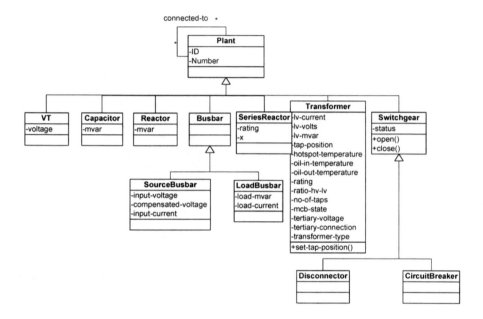

Fig. 6.5. Ontology of substation plant (uml class diagram), based on substation simulator data and transformer monitoring system

Information Management Ontology

The information management ontology, shown in Figure 6.6, provides the predicates used for document retrieval and querying. The properties of the *resource-description* class are based on the Dublin Core metadata standard [124].

This is a very basic ontology, which would not be suitable for all information retrieval applications, but fits with the needs of this system. Each document is considered to be a *resource*, which is described by a *resource description*. Also, a query has a particular *relevance* to each resource, which is represented by a real number between 0 and 1. Relevance is defined as a three-place relation:

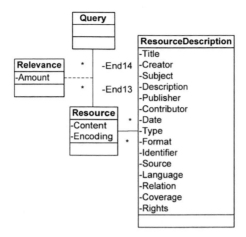

Fig. 6.6. Basic information systems ontology (uml class diagram)

$$relevance(Document, Query, Relevance)$$

Therefore, it is possible to query an information management agent, such as a document agent, using a query such as:

```
(all
    (sequence ?doc ?rel)
    (relevance
        ?doc
        (query :keywords
            (set ''transformer'' ''maintenance''))
        ?rel))
```

The above example means "Find all sets of a document *?doc* and relevancy *?rel* such that the relevance of *?doc* to the query "transformer maintenance" is *?rel*". In the example above, the query is represented as a frame containing a set of keywords. In the current implementation, this is abbreviated to just the set of keywords, as no other type of query is permitted.

6.4 Examples of Usage

This section provides several examples of how the system may be used to perform various tasks, including data querying, mobile agent data analysis and remote operation. Example FIPA ACL messages are provided to demonstrate the use of the FIPA standard protocols. Each ACL message has a sender, receiver, content (which is written in the FIPA SL) and protocol. In these examples, the agent name **user@pc2214:1099/JADE** represents the user agent.

Certain message parameters (*e.g.* conversation-id, ontology) have been omitted from the messages for brevity. Also, searches of the DF are not shown. Each time an agent wishes to carry out an action or to submit a query, it will first search the DF to find other agents capable of processing that request or answering that query, unless an appropriate agent is already known and is still available (has not disconnected). This means that it is possible to substitute different agents providing the same information, and permits agents to be added and removed at runtime. In these examples, code appears in typewriter font (*e.g.* `code`) and comments in roman font.

6.4.1 Querying IMU for a Data Set

In this system, a *data set* represents a series of events, each specifying the value of a property of an item of plant at a certain time. The user agent provides the user with the ability to define a data set to be retrieved using the name of the property and item of plant, a start time and an end time. In the future, additional criteria to define the data set could be added by modifying the user agent and graphical user interface.

Step 1: Select Plant Class

In order to assist the user in generating a query, the user agent provides a series of steps for query generation. The agent first retrieves the names of all plant classes (those that are a subclass of *plant*) from the ontology agent. This is done using the following message (assuming that the name of the ontology agent is `onto@pc2214:1099/JADE`):

```
(query-ref                                A query message
    :sender
        (agent-identifier                 Sent by user agent
            :name user@pc2214:1099/JADE)
    :receiver
        (set
            (agent-identifier             To ontology agent
                :name onto@pc2214:1099/JADE))
    :content                              Query to find all plant
        ''((all ?a (subclass-of ?a plant)))''
    :protocol fipa-query)                 FIPA Query proto-
                                          col
```

The ontology agent should then reply, providing the names of these classes. For example, suppose that the available classes are *transformer, circuit-breaker and switchgear*:

```
(inform                                   Information message
    :sender
        (agent-identifier                 From ontology agent
```

```
            :name onto@pc2214:1099/JADE)
        :receiver
            (set
                (agent-identifier              To user agent
                    :name user@pc2214:1099/JADE))
        :content
            ''(= (all ?a (subclass-of ?a plant))   Query results
                (set transformer circuit-breaker switchgear))''
        :protocol fipa-query)               FIPA Query proto-
                                            col
```

The user agent then presents these classes to the user as a list, and the user
selects a class in which they are interested. The system may then proceed to
the next step.

Step 2: Select Individual Item of Plant

Once the user has selected a class of plant, the user agent will then retrieve the
instances of that class from the static database agent. For example, suppose
that the *transformer* class has been selected:

```
    (query-ref                              Query message
        :sender
            (agent-identifier               From user agent
                :name user@pc2214:1099/JADE)
        :receiver
            (set
                (agent-identifier           To ontology agent
                    :name onto@pc2214:1099/JADE))
        :content                            Find all transformers
            ''((all ?a (instance-of ?a transformer)))''
        :protocol fipa-query)               FIPA Query proto-
                                            col
```

Now suppose that the available instances of *transformer* are *sgt1, sgt2* and
sgt3:

```
    (inform                                 Information message
        :sender
            (agent-identifier               From ontology agent
                :name onto@pc2214:1099/JADE)
        :receiver
            (set
                (agent-identifier           To user agent
                    :name user@pc2214:1099/JADE))
        :content                            Query results
            ''((= (all ?a (instance-of ?a transformer))
                    (set sgt1 sgt2 sgt3)))''
        :protocol fipa-query)               FIPA Query proto-
                                            col
```

Step 3: Select Property

The user agent must now determine the properties of the selected object. This is a more complex operation than the previous two. The ontology agent holds information about the properties of classes, while the static database agent holds information about which classes a particular object belongs to. It is not sufficient in all cases simply to query the ontology agent for the properties of the class selected previously, because if the object is actually an instance of a subclass of the selected class, there may be properties of the object that are not properties of the selected class. Therefore, this step involves the integration of information from the two databases. This can be done either by the user agent or by the database agents; which of these alternatives is best is discussed in Section 7.2.2. Here we suppose that the integration is done by the database agents, and that the user agent only queries the static database agent[2].

Supposing that the object selected by the user in the previous step was *sgt1*, the following query will be sent to the static database agent (static@pc2214:1099/JADE) (note that a *slot* is equivalent to a property, but is the term used by the FIPA Ontology service):

```
(query-ref                                   Query message
    :sender
        (agent-identifier                    From user agent
            :name user@pc2214:1099/JADE)
    :receiver
        (set
            (agent-identifier                To static DB agent
                :name static@pc2214:1099/JADE))
    :content ''((all ?a (slot-of ?a sgt1)))''  Find all slots of SGT1
    :protocol fipa-query)                    FIPA Query proto-
                                             col
```

The static database agent must then do two things: retrieve the class of sgt1 from its own database, and then retrieve the properties of that class from the ontology agent. Having determined that sgt1 is an instance of *transformer*, the static database agent sends the following message to the ontology agent[3] :

```
(query-ref                                   Query message
    :sender
        (agent-identifier                    From sta-
                                             tic DB agent
```

---

[2] This is how this is currently accomplished in the prototype system.

[3] In an optimal implementation, this is the message that will be sent. In practice, because of the *ad hoc* implementation of distributed backtracking in the prototype which does not propagate the *any* or *all* quantifier from the originating query, a series of query-ref messages will be used, containing *any* queries rather than *all* queries.

```
                    :name static@pc2214:1099/JADE)
          :receiver
              (set
                  (agent-identifier                To ontology agent
                      :name onto@pc2214:1099/JADE))
          :content
                  ''((all ?a (template-slot-of ?a transformer)))''
          :protocol fipa-query)                    FIPA Query proto-
                                                   col
```

The ontology agent will then reply with the properties of the *transformer* class. Suppose that these are *lv-current, lv-mvar* and *lv-volts:*

```
(inform                                        Information message
      :sender
          (agent-identifier                    From ontology agent
              :name onto@pc2214:1099/JADE)
      :receiver
          (set
              (agent-identifier                To static DB agent
                  :name static@pc2214:1099/JADE))
      :content                                 Query results
          ''((= (all ?a (template-slot-of ?a transformer))
                    (set lv-current lv-mvar lv-volts)))''
      :protocol fipa-query)⁴                    FIPA Query protocol
```

Finally, the static database agent will forward the slot names to the user agent:

```
(inform                                        Information message
      :sender
          (agent-identifier                    From sta-
                                               tic DB agent
                  :name static@pc2214:1099/JADE)
      :receiver
          (set
              (agent-identifier                To user agent
                  :name user@pc2214:1099/JADE))
      :content
              ''((= (all ?a (slot-of ?a sgt1))   Query results
                      (set lv-current lv-mvar lv-volts)))''
      :protocol fipa-query)                    FIPA Query proto-
                                               col
```

---

[4] As for the query message, in the current implementation there will be many of these messages. It is hoped that this issue may be resolved in a later implementation.

Step 4: Retrieve Data Set

Once the user has selected the object, property, start time and end time of the data set, this information is converted into a FIPA ACL query and forwarded to the IMU agent. For example, suppose that the user has selected *sgt1* as the object, *lv-current* as the property, 13/3/03 12:00:00 as the start time and 15/3/03 12:00:00 as the end time. The user agent will then query the IMU agent with the following message:

```
(query-ref                                    Query message
   :sender
      (agent-identifier                        From user agent
         :name user@pc2214:1099/JADE)
   :receiver
      (set
         (agent-identifier                     To ontology agent
            :name onto@pc2214:1099/JADE))
   :content ''((all (set ?a ?t)    Query for current values be-
                                   tween specified times
                 (and
                    (t (lv-current sgt1 ?a) ?t)
                    (and
                       (?t > 13032003T120000000)
                       (?t < 15032003T120000000)))))''
   :protocol fipa-query)                       FIPA Query proto-
                                               col
```

The IMU agent will then (using the Prolog rules of its reasoning engine) convert this into a call to the IMU web service to retrieve the relevant data, which will be passed back to the user interface agent. The user agent then displays the data on the screen as a graph and table.

Discussion

The database querying procedure functions correctly in the prototype implementation, and is relatively simple to use. The main problem lies in the performance of this process when handling large data sets. This is due to both the method of retrieving data from the database using backward chaining, which means that only one row of data at a time is retrieved, and the use of the string-based SL language, which generates large messages in comparison to binary encodings and requires computing time to be expended in constructing and parsing the messages. It can be seen from the mobile agent data analysis benchmark in Section 5.2.2 that the wrapper-based mobile agent (which used the database agent described here) performed approximately 3.4 times worse than a mobile agent with direct database access and data caching (probably close to the optimal solution). However, this does not show how much of this performance difference is due to data retrieval and how much to the use of string-based messaging. The problem of slow data retrieval might be alleviated by using a different implementation of the server agent, which could retrieve multiple rows at a time. However, all string-based data representations, including the commonly used XML format, encounter performance problems, and so these would be more difficult to solve. For example, [156] describes problems caused by the XML format generating large data files and creating overheads in terms of parsing and transformation. The only solution to this problem would appear to be the use of binary messages, which would remove the advantages of using an explicit, standardised and implementation-independent knowledge representation.

6.4.2 Mobile Agent-based Analysis of Data

The mobile agent-based data analysis proceeds in three steps. First, the data set or data sets to be analysed must be defined. The configuration file for the mobile agent is then generated by the user interface agent, and the mobile agent is launched. Finally, the mobile agent carries out the analysis and displays the results to the user.

Step 1: Define Data Sets and Analysis Report

For mobile agent-based data analysis, the steps used to define a data set are the same as those for a database query, except that a mobile agent is capable of analysing multiple data sets. Therefore, the graphical user interface used is the same except for buttons allowing the user to move between data sets. Each data set is defined as described in Section 6.4.1, Steps 1–3.

Once the data sets have been selected, the user must specify the analyses to perform. This is done by selecting an analysis function, and then specifying the data set on which this function will operate, and the element of the data set that will be used (either the plant property or the time). Some analysis functions may operate on multiple arguments, which may be from the same or different data sets. The user then specifies the report by entering any text which should appear with the analysis results.

Step 2: Launch Mobile Agent

Once the user has entered all of the details required to perform the analysis procedure, the user agent generates a configuration file and launches the mobile agent described in Section 5.2. When started, the agent reads the configuration file. It then retrieves each data set in turn, carrying out each analysis as soon as all of its required data sets have been retrieved. The messages used to retrieve the data set are the same as those described in Section 6.4.1, Step 4, except that the role played by the user agent is played by the mobile agent. Once all analyses have been performed, the report is generated.

Step 3: Display Results

The results of the analysis are displayed by the mobile agent using the system's default Web browser, as shown in Figure 6.7. The large gap between the sequence of data points to the left of the graph and the single point to the right is due to a discontinuity in the example data being used.

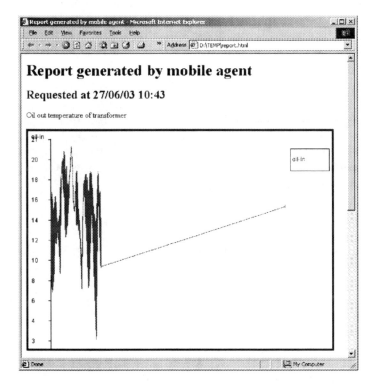

Fig. 6.7. Report generated by mobile agent

6.4.3 Searching for Documents

As only a single document agent is used in the prototype system, there is no need to attempt the integration of results from multiple sources, which is a difficult problem for which research is still ongoing [136, 157, 158]. To search for documents, the user first inputs a set of keywords into the HMI (the text entry box marked by "A" on Figure 6.8)

These are then transmitted to the user agent via the DataSocket connection. The user agent then sends a message to the document agent, requesting it to inform the user agent of all documents relevant to that query. For example, suppose that the query chosen by the user was "transformer monitoring". The following message might then be sent:

```
(request                                    Request message
    :sender
        (agent-identifier                   From user agent
            :name user@pc2214:1099/JADE)
    :receiver
        (set
            (agent-identifier               To document agent
                :name doc@pc2214:1099/JADE))
    :content ''((all                        Query for documents
                (sequence ?d ?r)            relevant to
                (relevance                  transformer monitor-
                                            ing
                ?d
                (set ''transformer''
                     ''monitoring'')
                ?r)))''
    :protocol fipa-query)                   FIPA Query protocol
```

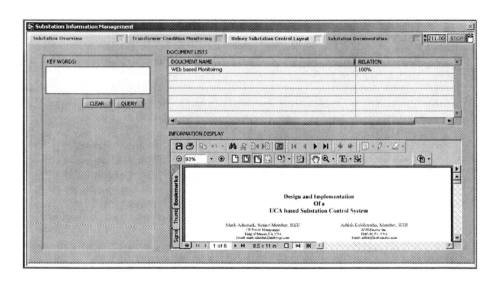

Fig. 6.8. User interface for document querying

This would be followed by a reply from the document agent, giving the resource descriptions of any relevant documents. The names and relevancies of these documents are passed from the user agent to the user interface and displayed in the list box marked "B" on Figure 6.8.

```
(inform                                     Information message
    :sender
        (agent-identifier                   From document agent
            :name doc@pc2214:1099/JADE)
    :receiver
        (set
            (agent-identifier               To user agent
                :name user@pc2214:1099/JADE))
    :content ''((=
                (all                         Query
                    (sequence ?d ?r)
                    (relevance
                        ?d
                        (set ''transformer''
                             ''monitoring'')
                        ?r)))
                (set                         Query results
                    (sequence
                        (resource-description
                            :title iSCSBrA4)
                        0.9705195)
                    (sequence
                        (resource-description
                            :title ''SS7 - Bricker'')
                        0.95818204))
            ))''
    :protocol fipa-query)                    FIPA Query protocol
```

The user agent may then request the document agent to transmit the contents
of a document to it. These will be encoded using the Base64 encoding. The
protocol used is *fipa-request*, and the message format is [5]:

```
(request                                     Request message
    :sender <user agent>                     From user agent
    :receiver (set <document agent>)         To document agent
    :content ''((action <document agent>     Retrieve documents
                (retrieve s<resource description>)))''
    :protocol fipa-query)                    FIPA Query protocol
```

The document agent then sends an *agree* message to the user agent, as specified by the fipa-request protocol, and following that encodes the document
and transmits it to the user agent as an *inform* message[6]:

---

[5] where < user agent> is replaced by the agent identifier of the user agent, <
document agent> by the agent identifier of the document agent and < resource
description> by the resource description of the document required.

[6] < action> is the action expression contained in the previous message.< docu-
ment> represents the encoded document content. < format> is the format of the
file, expressed as a MIME type (*e.g.* "application/pdf").

```
(inform                                    Information message
    :sender <document agent>               From document agent
    :receiver (set <user agent>)           To user agent
    :content ''((result                    Set of documents
                <action>
                (resource
                    :content <document>
                    :encoding base64
                    :format <format>
                    :title <title>
                )))''
```

6.4.4 Performing an Action Using Data Acquisition Agents

To set the value of a property of an item of plant, the *fipa-request* protocol
is used. The user agent must first send a *request* message to the appropriate
plant agent (if the appropriate agent is unknown, it can be located using the
DF). For example, suppose that the user wishes to open a circuit breaker
"x10", and that the name of the plant agent is x10@pc2214:1099/JADE. The
user first sends a request to the user agent via the graphical user interface
shown in Figure 6.9.

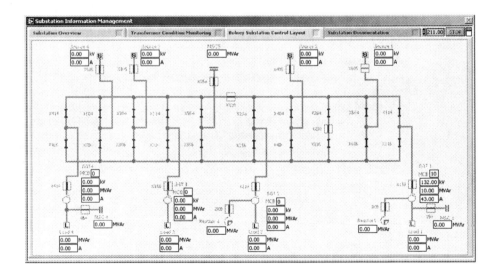

Fig. 6.9. Graphical user interface for substation control

The user agent then sends the following message to the plant agent (there
may also be other message parameters used such as *conversation-id* or *reply-
with* if the user agent wishes to track the conversation using these parameters):

```
(request                                      Request message
    :sender
        (agent-identifier                     From user agent
            :name user@pc2214:1099/JADE)
    :receiver
        (set
            (agent-identifier                 To breaker agent
                :name x10@pc2214:1099/JADE))
    :content
        ''((action
                (agent-identifier             Open breaker
                    :name x10@pc2214:1099/JADE)
                (open x10)))''
    :protocol fipa-request)                   FIPA Request protocol
```

The plant agent will then consult its mapping rules to determine the data acquisition node, device and channel responsible for the control of x10, and the value to be written to that channel which will result in x10 being opened. Supposing that the channel used is DC1, and the device agent is device@pc2214:1099/JADE:

```
(request                                      Request message
    :sender
        (agent-identifier                     From breaker agent
            :name x10@pc2214:1099/JADE)
    :receiver
        (set
            (agent-identifier                 To device agent
                :name device@pc2214:1099/JADE))
    :content
        ''((action
                (agent-identifier             Write value to channel
                    :name device@pc2214:1099/JADE)
                (write-value dc1 0)))''
    :protocol fipa-request)                   FIPA Request protocol
```

The device agent will then carry out the request, and send an *inform* message to the plant agent to notify it that the action is complete. The plant agent will then notify the user agent in the same way. To do this, the fipa-request protocol is used.

6.4.5 Reading a Plant Property Using Data Acquisition Agents

To read a property, the *fipa-query* protocol is used. The user agent first locates the appropriate plant agent. Then the user agent sends a *query-if* or *query-ref* message to that agent (depending on whether the user wishes to confirm the value of a property or to find out the value of that property). For example, suppose that the user wishes to determine the *lv-current* (Low

voltage current) of transformer *sgt1*, and that the name of the plant agent is
sgt1@pc2214:1099/JADE. The following message would then be used:

```
(query-ref                              Query message
    :sender
        (agent-identifier               From user agent
            :name user@pc2214:1099/JADE)
    :receiver
        (set
            (agent-identifier           To transformer agent
                :name sgt1@pc2214:1099/JADE))
    :content ''((iota ?a (lv-current sgt1 ?a)))''   Get cur-
                                        rent value
    :protocol fipa-query)               FIPA Query protocol
```

The plant agent must then use its mapping rules to locate the appropriate
device agent and channel. Suppose that the channel name is SL1, and the
device agent is device@pc2214:1099/JADE. The plant agent will then send
the following message:

```
(query-ref                              Query message
    :sender
        (agent-identifier               From trans-
                                        former agent
                :name sgt1@pc2214:1099/JADE)
    :receiver
        (set
            (agent-identifier           To device agent
                :name device@pc2214:1099/JADE))
    :content ''((iota ?a (value sl1 ?a)))''   Get value of channel
    :protocol fipa-query)               FIPA Query protocol
```

The device agent will then read the value of the SL1 channel, and send a reply
to the plant agent. Supposing that the value is 12.5:

```
(inform                                 Information message
    :sender
        (agent-identifier               From device agent
            :name device@pc2214:1099/JADE)
    :receiver
        (set
            (agent-identifier           To transformer agent
                :name sgt1@pc2214:1099/JADE))
    :content                            Value of channel
        ''((= (iota ?a (value sl1 ?a)) 12.5))''
    :protocol fipa-query)               FIPA Query protocol
```

Finally, the plant agent sends a similar *inform* message to the user interface
agent.

6.4.6 Human—Machine Interaction Using Personal Assistant Agents

The PAA, $agent_{PAA}$, is implemented using Component Object Model (COM) techniques. The services for substation information management that are usually provided by $agent_{PAA}$ are shown in Figure 6.10, in which a number of functions of the system operation and management are included. The roles defined in the agent community and their distribution to HMI depend on additional functional specifications of the system. There are two types of user service roles which provide personal services:

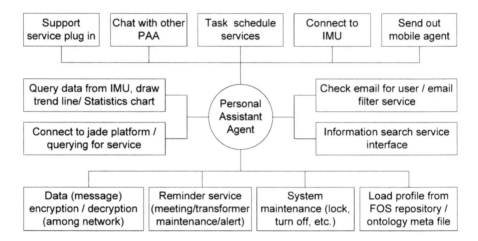

Fig. 6.10. Services of the personal assistant agent

1. *User interface management role*: a user friendly interface which interacts with users and enables users to express their requests. The role can refine the request based on the ontology specification and present the results in a graphical format to the user.
2. *User profile management role*: a role that is responsible for the management of a user's profile. A user profile agent can be regarded as an archivist. The annotations of archives can be any ontology entities about user and user agents, for instance, a communication log of the $agent_{PAA}$ and a personal profile of a user, *etc*. The user profile management role enables the profile to be used not only for interface purposes, but also for learning techniques and proactive searches, *etc*.

These two recurrent roles in the user service community may be merged into one agent or more agents which may be jointly performed to implement specific functionalities of the system. Two functionalities are expected for the management of users, one is the management of an interface that enables the interaction between the user and the multi-agent system; the other is updating (*i.e.* customising and learning) and exploiting the user's preferences while the user interacts with the $agent_{PAA}$.

The Agent Profile

The $agent_{PAA}$ learns the user's preference by daily interactions. It also provides a user profile input interface. The configuration pages give a user-friendly way to define some particular information sources and task schedules for displaying user profile, illustrating the services which are available for the user and correcting the IMU configuration information. These functions are shown in Figure 6.11.

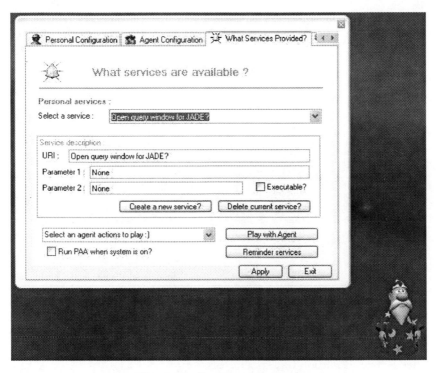

Fig. 6.11. Services registration

Furthermore, the information of the agents which registered on the agent platform will be displayed by $agent_{PAA}$, which provides the capability to subscribe information services.

Remote Control

The $agent_{PAA}$ provides the capability of remote hardware control. There is an example of remote relaying control of a switcher "x605", in a NGT substation, using $agent_{PAA}$, as shown in Figure 6.12. Four steps are taken in this process.

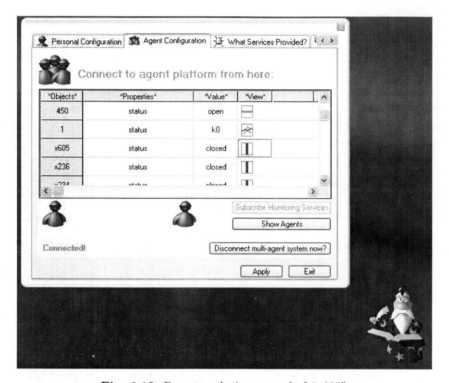

Fig. 6.12. Remote relaying control of "x605"

1. The $agent_{PAA}$ subscribes substation condition monitoring services. States of facilities at substation are then updated according to the changing events, which is informed by agents who own the information resources.
2. "Switch off the targeted switcher" is commanded by user. The $agent_{PAA}$ needs a confirmation to perform this action.
3. The $agent_{PAA}$ consults the DF agent $agent_{DF}$ for service descriptions. $agent_{DF}$ replies with matched agent identifiers. "agent x605@pc032070:1099 /JADE can do this job", and the $agent_{x605}$ agent identifier has been confirmed in this case.
4. The $agent_{x605}$ performs switch off action and informs the closure state of the switcher "x605".

The states of the switcher "x605" are listed in Figure 6.12.

Information Services

Ontology-driven Information Query

An example of ontology-driven information query is given in Figures 6.13 and 6.14.

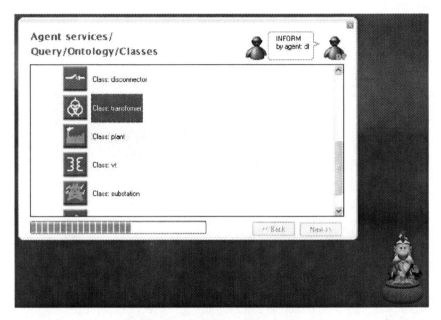

Fig. 6.13. Concepts of power system

Figure 6.13 displays the components of a power system, such as *"disconnecter"*, *"transformer"*, *"plant"* and *"substation"*, etc. In this case, *"transformers"* situated in the power system are being toggled, which are named as *"SGT1"*, *"SGT2"*, *"SGT3""* and *"SGT4"*. The detail of each parameter of a transformer can be queried using $agent_{PAA}$.

Information Management Unit Access

The IMU defines the data structures according to the different disciplines and application domains. There is a standard database behind IMU. No third parties can access this database directly. A Simple Object Access Protocol (SOAP) is provided to unify the access interface. The IMU provides means to store substation historical data, which is provided by a data acquisition system connected to the substation simulator. However, the IMU does not provide control functionality, or publish—subscribe information update. Therefore, the $agent_{PAA}$ provides comprehensive IMU access services, which include:

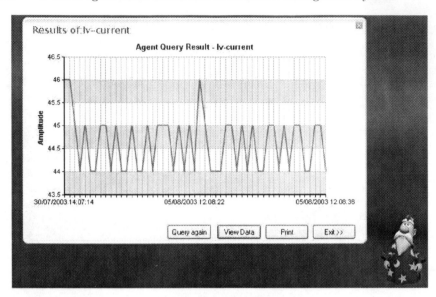

Fig. 6.14. Chart display

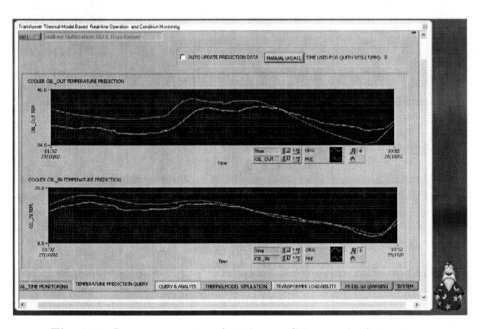

Fig. 6.15. Remote access to substation condition monitoring system

1. Historical data query. For example, the user selects the query object constraint "AMBT_ORG", which is the ambient temperature measurement, and time constraint, then presses "Query".

2. Report generation. The query results report will be generated by the $agent_{PAA}$.
3. "Real-time" monitoring: the $agent_{PAA}$ can keep monitoring a certain object by checking the state of this object continually.

Figure 6.14 displays the results of querying "*lv-current*" from a transformer.

Other Services

The $agent_{PAA}$ provides the capability to extend services by updating components, which is based on the COM technique, such as the documentation management and query service; thermal model based transformer condition monitoring report service, which includes all thermal model analysis modules; online Dissolved Gas Analysis (DGA) chart which portrays the individual concentrations of gases and the remote access of the substation transformer condition monitoring system. Figures 6.15 and 6.16 give two examples of the remote access to the substation condition monitoring system and the online DGA chart.

6.5 Implementation Issues

Initially it was intended to use the Information Management Unit (IMU) to provide the sole interface between the information management system and the substation simulator. However, this proved impossible because of the fact that the IMU did not provide "subscription" functionality—that is, it was not possible to have the IMU automatically update its agent when new information arrived. This meant that event updates had to be handled by periodically "polling" the IMU to retrieve new data, which is a relatively inefficient mechanism.

Introducing the data acquisition multi-agent system corrected this problem by providing direct access between the multi-agent system and the data acquisition device. However, this might cause problems in an operational deployment if, for security or other reasons, it was decided not to allow agents to access devices. This would mean that event functionality would either have to be implemented by polling or omitted from the deployed system.

Agents that require to be updated when new information becomes available in the system (for example, the plant agent needs to know what information is available from device agents regarding its item of plant) should do so by establishing a subscription with the DF to be informed of agents joining or leaving the system. This is not possible with the current system implementation but is available in newer versions of the JADE toolkit.

To allow agents not based on JADE (such as the user agent) to locate the agent platform, a basic broadcast discovery mechanism based on UDP is used, as none is provided by FIPA. However, as the FIPA work on ad hoc platforms

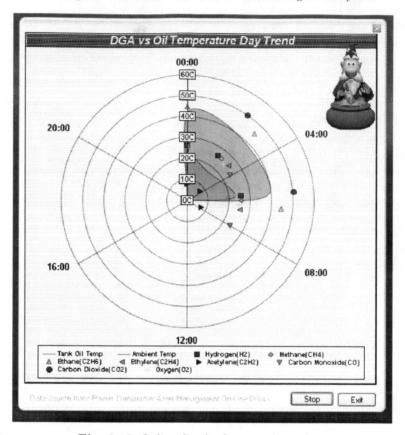

Fig. 6.16. Online dissolved gas analysis chart

progresses[7], it is possible that such a mechanism will be provided by FIPA platforms.

6.6 Summary

This chapter has described a prototype system based on the architecture of Chapter 3. The various agents described in Chapter 4 are implemented using the JADE platform, and a reasoning engine and knowledge base based on Prolog. Use of the Prolog language provides relatively simple conversion to and from FIPA SL, as both are based on first-order logic. The prototype demonstrates the feasibility of the architecture and shows how it may be implemented. It also reveals some of the problems involved in this implementation. For example, integrating the multi-agent system with the HMI

---

[7] www.fipa.org/activities/ad_hoc.html

platform written in LabVIEW^{TM} was a particularly time-consuming task, involving the implementation of a user agent in VisualBasic® (in order to access ActiveX® controls) and linking this agent to the JADE platform using a TCP/IP connection.

The next chapter presents an evaluation of the architecture, using both theory and the experience provided by the implementation of this prototype. It examines whether the architecture provides the required functionality, and considers its ability to be modified easily in response to changes in the substation plant or in the data acquisition system.

7

Evaluation and Analysis

This chapter presents a brief evaluation of the architecture described in Chapters 3 and 4. Where applicable, experience from the prototype described in Chapter 5 is used. The software engineering quality attributes of performance, modifiability, availability and security [159] are considered with reference to the described architecture. Different authors in the software engineering literature use different variations on these quality attributes, for example, Sjeko [160] uses a more complex quality model consisting of dependability (including safety and security), satisfaction, functionality, flexibility (including modifiability) and performance. The criteria used here are drawn from both of these sources.

It is very difficult to evaluate the information management functionality of the system. The system's document retrieval function is based on standard algorithms, and therefore the use of information retrieval metrics such as precision and recall would test only these algorithms and not the architecture itself. It can be shown that the database retrieval function is capable of retrieving data from a database. However, the data integration functions are largely based on previous work in other domains such as Infomaster™ [82] and RETSINA [80], and evaluating data retrieval performance would largely test the implementation, which is not of production quality, and not the architecture. The preferred method of evaluation would be to install the system in a substation and request that the substation engineers compare it with other systems and existing technology. However, the time that would be required to implement the system to a standard suitable for such a test meant that this was not possible. Therefore, this chapter presents a largely theoretical evaluation based on software engineering principles. This is sufficient to draw certain limited conclusions about the flexibility and modifiability of the architecture. Further research should involve comparison with other systems and user-oriented testing.

7.1 Functionality

The functionality provided by a system implemented using the proposed architecture should, at least, be capable of matching that provided by a traditional, object-oriented or Web-based HMI/SCADA substation automation system. Various criteria taken from the literature are now used to describe the desired functionality and compare it to that provided by the prototype system. However, it is important to note that the work described in this book has developed an architecture for the construction of industrial automation systems, rather than a specific automation system. The prototype is intended only to demonstrate certain functionalities of the architecture, and not to include every feature of a full substation automation system. Therefore, many of the features described in these criteria have not been implemented. Where this is the case, the way in which a feature might be implemented using the multi-agent system is discussed.

7.1.1 National Grid Transco Requirements for Substation Control Systems

The NGT requires a substation control system to be capable of fifteen functions [116]. For each of these functions, Table 7.1 considers how they might be implemented using the multi-agent architecture described in this book.

Therefore, the current system provides only a subset of the functionality provided by a substation control system. However, the focus of this research has been on the design of a generic architecture, rather than the implementation of a complete system. Therefore, a number of functions are abstracted in the architecture as generic "task-specific agents".

7.1.2 Haacke "Opportunity Matrix"

Haacke *et al.* [161] provide a list of "candidate functions" for a substation automation system which aim to address the needs of a power operating company, and are not (in their view) currently being met. These are as follows:

1. Equipment condition monitoring
2. Automatic load restoration
3. Dynamic transformer ratings
4. Adaptive relay settings for distribution circuits
5. Power system disturbance and power quality data
6. Feeder automation support
7. Expert alarm processing
8. Access to substation metering data
9. Access to power company documents and systems
10. Corporate data repository
11. Additional SCADA quantities

Table 7.1. National Grid Tansco requirements

Requirement	Implementation using multi-agent system
Local control	Performed using the operator intervention process through HMI, user interface agent, plant agents, device agents and IEDs.
Telecontrol	Performed using the same process as local control. Mobile agents may also be used for predetermined sequences of interactions.
Alarm annunciation	May be performed by a task agent using data supplied by plant agents to generate alarms.
Data archiving	Uses data logging database or Information Management Unit.
Synchronisation Plant Performance Monitoring Delayed Auto Reclose Switching Automatic Tap Change Control Automatic Reactive Switching Fault Recording Primary System Monitoring Sequential Isolation Switching Interlocking	These functions are not implemented in the prototype. In a full system, they would be implemented either using task agents (one agent for each task) or by modifications to the control rules of the plant agents.
Database Creation and Amendment	Performed by data storage agents.
Diagnostic Facilities	This function is not currently implemented and should be considered in further work.

12. Adding SCADA to non-SCADA substations
13. Training simulator

Of these suggested needs, the multi-agent architecture may contribute to three:

- Access to documents and systems (9): The use of a multi-agent architecture and IP networks allows access to external systems, via the use of wrappers. For example, the document agent provides access to documents. However, it would also be possible to implement this function using a client–server or distributed object approach.
- Data repository (10): The multi-agent system maintains individual data repositories in each substation. However, the use of mobile agents provides a capability to integrate data from different databases, hence providing a type of "distributed data repository".
- Additional SCADA quantities (11): By adding additional provider agents to the system, additional quantities may be made available at runtime.

Additionally, equipment condition monitoring (1) and expert alarm process-ing (7) might be implemented using a multi-agent methodology, but are not currently part of the system or architecture described in this book.

7.1.3 Summary of Functionality Results

Overall, the multi-agent system, extended with additional agents, should prove capable of providing all of the functionality provided by a traditional automa-tion system. However, at least without the addition of artificial intelligence, it does not offer any significant additional functionality that cannot be imple-mented by other means, although, as stated by [57] for mobile agents, it does provide a consistent framework for the implementation of different functions and tasks, such as control and information management. In order to be a use-ful technology, the multi-agent system should also provide advantages in other areas, such as performance, modifiability or ease of system development.

7.2 Performance

It has been shown in Chapter 5 that, in certain circumstances, mobile agents provide increased performance over client−server alternatives. Therefore the use of mobile agents is not discussed here. However, other aspects of the architecture's performance are considered. First, we consider the performance of the architecture for data acquisition and control tasks. We then consider the response time of the system, when performing information management tasks, to user queries.

7.2.1 Data Acquisition Performance

Because of the two-layered structure of the data acquisition system, shown in Figure 7.1, when new data arrives it must first be acquired by the device agent, and then passed to the relevant control agent. This means that, for control applications in which a control agent must respond to events, it is necessary for the data to pass through at least one intermediate agent (the device agent) before reaching the control agent. This could be a problem if it is necessary to achieve very fast control.

An alternative structure, shown in Figure 7.2, would be for the control agents to retrieve information directly from the data acquisition devices, using device-specific communications protocols or APIs. However, this compromises the modularity and information abstraction provided by the separate device agents and control agents, as it requires a control agent to have knowledge regarding a number of (possibly heterogeneous) data acquisition devices, and a number of separate capabilities and protocol drivers for accessing these devices.

The performance of different configurations of device agent and control agent is now analysed in greater detail.

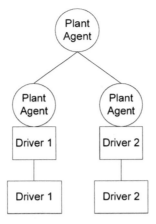

Fig. 7.1. Plant agent and device agents

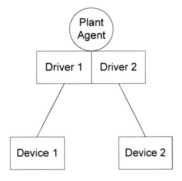

Fig. 7.2. Plant agent with direct access to devices

Message Passing

Suppose that a plant agent a wishes to send a command cmd to a device d over a network of bandwidth τ and latency δ, and receive a reply ack when the command is complete. The procedure taken is as follows:

1. Plant agent sends message to device
2. Device carries out action
3. Device sends acknowledgment to plant agent

Now suppose that instead of transmitting the command directly to the device, the plant agent communicates first with an intermediate device agent da. In this case, the following procedure must be carried out:

1. Plant agent sends message to device agent

2. Device agent processes message
3. Device agent sends message to device
4. Device carries out action
5. Device sends acknowledgment to device agent
6. Device agent processes acknowledgment
7. Device agent sends acknowledgment to plant agent

In the first scenario, two messages are sent (a to d, d to a). In the second scenario, four messages are sent (a to da, da to d, d to da, da to a). For each message, there is a cost to encode the message in a particular protocol (either a device protocol or an ACL) and a cost to interpret the message, along with the cost of transmitting it across the network. However, it is not necessarily the case that all messages must be sent across the network. Whether this is the case for different configurations of plant agents and device agents will now be examined.

Co-location Scenarios

In order to determine the number of messages transmitted across the network, it is necessary to determine whether any of the three objects involved (device agent, control agent and device) are co-located. In the general case, if we accept that each item of plant may be managed by multiple devices and each device may manage multiple items of plant, it is not possible for all three to be co-located. Suppose agent a manages item of plant p_1, which is managed by devices d_1 and d_2. If these two devices are not at the same node ($location(d_1) \neq location(d_2)$), it is impossible for a to be co-located with both.

Although, conceptually, it would be preferable to locate the device agent on the device, it would be possible to co-locate the device agent with the control agent, although this would only be the case if either each device agent corresponds only to one control agent, or all control agents corresponding to a particular device agent may be located with that device agent (Figure 7.3).

In this case, only the messages between the device agents and their respective devices are transmitted across the network.

If, instead, da and d are co-located (Figure 7.4), the messages between the plant agent and the device agents are transmitted across the network, whereas the messages between the device agents and the devices are transmitted locally. The number of messages transmitted across the network for a given interaction is the same in both cases. Therefore, any performance difference is due to the relative efficiency of the agent−agent and agent−device protocols. Also, it may not always be possible to locate an agent on a device.

Other Factors

Other factors affecting the performance of the data acquisition system include the relative inefficiency of general agent communication languages compared

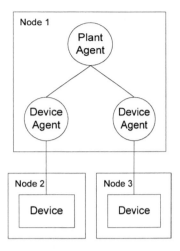

Fig. 7.3. Plant agent co-located with device agents

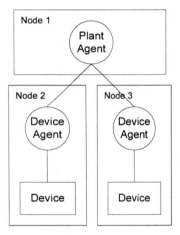

Fig. 7.4. Device agents co-located with devices

to specific protocols designed for data acquisition and control. This results in increased message size and in longer parsing times.

7.2.2 Responding to User Queries

The time taken to respond to a query from the user varies widely depending on the type and complexity of the query, and also on whether it relates only to one substation or to multiple substations. It also depends on the mechanism (mobile agent or static agent) used to answer the query, and whether information from multiple agents is required to produce an answer.

There are a number of possible bottlenecks involved in this procedure. The database wrapper agents might become overloaded if multiple queries were submitted simultaneously. Also, the directory facilitator might become overwhelmed by a large number of requests, although it might be possible to avoid this by introducing multiple federated DFs.

It has been observed that providing the database agents with the ability to interact as described in Section 4.1 produces a significant delay in query answering, due to the number of interactions between database agents involved in a typical backtracking procedure. It would be possible to reduce the load on the database agents by removing the ability for them to interact with each other, and having another agent (possibly the user agent or a broker) which integrated the information from multiple databases. However, the total number of agent interactions would not be reduced, and may in fact increase as the broker agent must interact with multiple database agents. An alternative strategy would be a modification of the inter-agent backtracking process, for example, to retrieve all the relevant results at once and then backtrack locally through the retrieved list. However, this would only reduce, rather than eliminate, the extra workload on the database agents.

7.2.3 Data Display

The National Grid specification for substation control systems [116] provides stringent performance requirements for the display of data in a substation control system, for example that data should be displayed on the VDU screen within a mean time of 2 seconds of a digital input changing state, with a standard deviation of 0.5 seconds. While it would not be possible to completely ensure that the system meets these requirements with a basic prototype implementation, it is necessary for the architecture to be designed in such a way that a more developed implementation would be able to do so.

The time to display data on the HMI of the system is affected by two things; the data acquisition speed, discussed in Section 7.2.1, and the amount of time taken for the user agent to retrieve data from the plant agents. The number of ACL messages involved in data display (at minimum) would be two:

1. Message from device agent to plant agent.
2. Message from plant agent to user agent.

This assumes that subscriptions are set up between user agent and plant agent and between plant agent and device agent. Providing that all agents are located in the substation, it will be possible in an otherwise unloaded network for these messages to be transmitted within the 2 second limit (in fact in a much shorter time-scale). For example, the time taken to transmit a 9905 byte ACL message between two containers on the same computer in the experiment documented in Section 5.3 was 8.05 milliseconds (although a longer time than this would be required to create the message and for the client to parse its contents). However, the effects of loading on the system have not been tested in the prototype, and remain an issue for further work.

7.3 Modifiability

One of the important claimed advantages of agent-based systems is their flexibility and ease of modification. For example, Ferber [8] states that multi-agent architectures are "especially suitable" for adapting to changes in the system in which it executes and to changing requirements.

7.3.1 Modifying the Substation

Replacement of Substation Plant

If an item of plant is replaced with a new one of the same type, then the details of the replacement item must be entered into the static database. If any of the monitoring devices related to the item of plant are changed, then the procedure described under "adding an IED" must also be carried out.

Because the plant agent obtains its plant configuration from the substation databases, it is not necessary to re-program this agent, but it must be restarted in order to force it to re-read the configuration. This could be done remotely using the administrative interface of the agent platform. However, it might also be possible for the plant agent to provide a command which could be sent in an ACL message and would cause it to restart itself and re-read its configuration.

Addition or Removal of Plant

If a new item of plant is added of an existing category (already in the ontology) then:

- The details of the item of plant must be added to the static database
- A plant agent must be created for that item of plant. It is likely that only minor changes to the configuration of the control agent would be required, as it could be modified from an existing plant agent for a similar item of plant.
- The mapping rules relating to the item of plant must be added to the mapping database.
- The user interface must be modified to display the new item of plant.

If an item of plant is added which is of a category not already present in the substation, then additionally the details of that category of plant must be added to the ontology.

Furthermore, when an item of plant is added it is necessary to add a number of IEDs to monitor that plant. Therefore, for each of these IEDs the procedure described in "Adding a new IED" must be carried out.

Adding a New IED

When a new IED is added, the following steps must be taken:

- Program a device agent for that IED, or modify an existing agent.
- Add the IED configuration to the static database.
- If the IED monitors a property of an item of plant which is not present in the ontology, then the ontology database will need to be updated. It may also be necessary to update the user interface to show the new property.
- Add the mapping rules for the IED to the mapping database.
- Restart the relevant plant agent to reload its mapping rules.

7.3.2 Modifying the Data Sources

Adding a New Database

To add a new database to the system, a wrapper agent for that database must be implemented. At minimum, this involves creating a set of rules to provide a mapping from the schema of the database into the global system ontology[1]. If the database is to act not only as a source of data, but is also to store data gathered by other agents in the system, it is also necessary for the database agent to establish subscriptions for appropriate data with the agents providing that data[2].

Adding a Document Repository

Providing that the documents in a document repository are of a format which can be handled by an existing document agent, the addition of a document repository to the system should involve only the instantiation of an agent to manage that document repository. No programming or compilation should be required.

However, supposing that a future incarnation of the system provides not only "search engine" functionality, but also information extraction, it would be

---

[1] This is not true in the case where the database uses an identical schema to another database already present in the system, in which case the mapping defined for that database can be re-used.

[2] It might also be possible to use two database agents—one reading and one writing. This might reduce the problems caused by load on either of the agents.

necessary to develop wrappers for the documents in the document repository to allow this extraction to be performed.

User and mobile agents will be able to locate the document repository automatically, without restarting, as the document agent will be registered with the directory facilitator[3].

Adding a New Type of Service

If a new service is to be made available of a type not already present, then the amount of work to be performed might be more substantial.

- If the service provided only query functionality, and the terms which could be queried were already available in the global ontology, then a wrapper agent would have to be written for the service. That wrapper agent would register with the directory facilitator, providing the list of available queries, and other agents would then automatically discover it and be able to query the service.
- If the service provides query functionality but additions to the global system ontology are required, the process is more complex. First, the global ontology must be updated to include these new terms. Any agents which need to use these terms may also have to be updated, unless the agent automatically discovers them from the ontology database, as is the case with the user interface agent (for mobile agent generation) and the plant agent (for information management only—if the information was to be used in control the control algorithms/rules of the plant agent would need to be altered as they would not take account of this information).
- If the service provides functionality other than simple data querying (for example prediction) then any client agents which need to make use of this service should be updated, as they will not be aware of the functionality that this service is capable of providing. The most likely agent to require alteration is the user agent.

7.3.3 Modifying the User Interface

Adding a New User

Adding a new user to the system would not currently require any changes to be made. However, in a system in which security was implemented it would be necessary to add any required details of the user (*e.g.* username and password) to the authentication service.

---

[3] This assumes that agents do not cache DF entries, and either search the DF each time they perform a document search, or subscribe to the DF for notification of new registrations (this is possible in certain versions of JADE and other platforms, but is not yet in the FIPA specifications).

Adding New Capabilities to the User Interface

To add capabilities to the user interface requires modifications to the user interface and possibly the user agent only, unless a new service is required that is not currently provided by the multi-agent system. For example, adding the ability to control the substation from the user interface in the prototype system required the addition of new input buttons, and the modification of the user agent (to read commands and transmit them to other agents) and the interface between the user agent and user interface.

7.3.4 Summary

As described above, the addition of a new item of plant or service requires relatively few changes in the multi-agent system. For example, alterations to an item of plant require only the device agent (and the configuration databases) to be modified. This is an advantage compared to a centralised system, in which modifications affect the entire system. However, the major drawback of the system described here is that many changes require alterations to the user interface or user agent (in order, for example, to display new items of plant on the one-line diagram) rather than just the agents directly responsible for an item of plant or database.

7.4 Security, Reliability and Availability

It is difficult to evaluate the security, reliability and availability of a prototype application. Therefore, it is necessary to consider similar results from other agent-based applications and architectures.

As far as security is concerned, the use of a static agent-based approach should not introduce additional problems beyond those introduced by a client−server model, providing that all agents in the system are developed and owned by a single entity (the power company). There are security and trust issues associated with *open* multi-agent systems, in which agents are developed by different entities; these issues are mentioned in [31].

Those security problems posed by the use of mobile agents should be relatively limited in the power systems domain, provided that all mobile agents originate within the company itself (the origin of a mobile agent could be verified using digital signatures as described in [61]). If it is not permissible for external users to submit mobile agents to the system, the problem of malicious mobile agents should be reduced. The problem could be reduced even further by preventing users from implementing their own mobile agents, and allowing them to use only the pre-implemented mobile analysis agent and mobile remote control agent. However, this would require that a large function library was available for the analysis agent, to provide all of the analysis functionality required by the different users. This process could be helped by extending the capability of the analysis agent to permit analysis functions to operate on the results of other analysis functions, rather than only on retrieved data sets. This would reduce the number of different functions required.

Fedoruk and Deters [162] suggest that the "brittleness" of multi-agent systems is a main contributor to their lack of deployment in industry, and that this brittleness is caused by the fact that there is no centralised control of a multi-agent system, and that therefore it is "difficult to detect and treat failures of individual agents". They found that introducing replicated agents significantly reduced the failure rate of a multi-agent system, but that this was at the price of increased system load. In a system such as a power system, in which reliability is important, it would probably be necessary to introduce replicated agents, possibly in addition to other techniques such as "watchdog" agents or agents which monitor each other's behaviour during task execution [163].

Failure of an Agent

One possible cause of failure in the system is the failure of one or more individual agents. If one agent establishes a subscription with another for some item of information, the subscriber will only receive messages from the provider when that item of information changes. Therefore, if the provider agent fails, the subscriber will not have any means to detect this, and will probably be unaware of any further changes to the item of information in question.

Another possible cause of failure is that the DF entries of agents remain after they have failed. This means that an agent may attempt to contact an agent that no longer exists, but which still has a DF entry. This problem is solved by the use of leasing (similar to that implemented by Jini^{TM}) in the latest version of the FIPA specifications.

7.5 Integration into Existing Substations

One measure of the practical usefulness of the architecture described in this book would be the ability to integrate it into existing substations which already have a substation control system or automation system.

The complete multi-agent system relies on access to the substation IEDs for monitoring and control. This means that for each IED, a protocol driver must be available. There would also be problems involving the development of task-oriented agents to replace functions of the substation control system, and the development of algorithms for control agent cooperation. Some projects are being developed by Intelligence Engineering and Automation Research Group, Department of Electrical Engineering and Electronics, University of Liverpool [164, 165]. Finally, there might be security concerns involved with the open architecture and the use of mobile agents.

However, it is possible to make more limited use of the multi-agent system to carry out only the task of monitoring and information management.

7.6 Possible Applicability to Other Industries

In the initial requirements for the architecture described in this book, it was desired that it should be as generic as possible, and able to be applied to a number of different industries. As a possible example, we consider the hospital ward described in [166]. In the hospital, there are seven wards, each with up to six babies. Each baby is connected to a monitoring system which continuously monitors a number of numerical parameters such as the electrocardiogram (ECG) waveform. The monitoring system also generates derived knowledge, such as the heart rate in beats per minute which is derived from the ECG. Each measured parameter is archived once per minute by an archival system [166, p2].

To adapt the architecture to this scenario, it is suggested that the plant agent be used to represent the baby, with a new configuration, set of control rules and ontology being defined for this purpose. The device agent would be used to represent the monitoring and control device.

The user interface agent should be able to be used without any modification, because the only ontological commitment it has to the substation domain is the use of the "plant" class, which could be kept as the root of the new domain ontology (covering humans). However, the HIM itself would have to be rewritten.

For the archival system, a database agent would be used and appropriate mapping rules would have to be created. This supposes that the archival system uses a standard database. If not, a special "archival agent" would have to be created (as for the IMU in Chapter 6), but it would perform similar functionality to a database agent.

One problem with the portability of the architecture in this scenario is that the term "plant" does not easily apply to a human, and, in order for the system to function correctly using its built-in ontology, it would be necessary to define "baby" as a subclass of "plant". However, the difficulties posed by this should be mostly cosmetic, and would possibly be hidden from the user by the HMI.

Another, more serious, problem, is that the measurements taken in the hospital scenario involve a high degree of uncertainty, which is not present in a power substation.

The conclusions drawn from this exercise have not been tested in practice. Therefore, during an actual implementation a number of difficulties might emerge which would necessitate changes to the architecture or to the individual agent implementations.

7.7 Discussion

From the evaluation in this chapter, a number of points can be made regarding the advantages and disadvantages of multi-agent systems in comparison to existing technologies for power system automation.

7.7.1 Advantages

Flexibility

As demonstrated in Section 7.3, the use of a multi-agent system, and in particular the directory facilities provided, enhances the flexibility of the system by permitting new devices and items of plant to be added without changing the software of the rest of the system. However, this advantage is reduced in the power systems domain by the fact that the system does not change rapidly (items of plant are rarely added to substations). Also, in the current prototype it is necessary to restart several of the agents (plant agents and device agents) in order to change their configuration. It would be useful to add a feature enabling the agent to read updated configuration rules at runtime.

The use of a multi-agent system provides a basis for the introduction of distributed control systems in which agents (perhaps representing items of plant or other entities) act in an autonomous manner without outside intervention. However, in the prototype system described here, agents do not exhibit significant autonomous behaviour, as it would be most useful in the automatic control task, which has not yet been implemented.

Inherent Distribution

The inherently distributed nature of the power system means that a multi-agent system, which provides autonomy to its constituent components, is well-suited to this application domain. For example, the use of agents to represent objects such as transformers and circuit breakers is a natural "fit" to the system being controlled. However, a similar structure might also be obtained using a distributed object system.

Integration

Using a multi-agent system provides a convenient framework to represent different tasks and to integrate different data sources. Rather than a number of separate software programs, all tasks are performed through the multi-agent system, enabling data to be shared between tasks. The use of a standard agent communication language provides a fixed communications mechanism which can be used by heterogeneous agents. However, there are other methods (such as distributed object systems) which might be able to achieve the same goals, although distributed object communications do not have the high-level semantic content of agent communication languages.

7.7.2 Disadvantages

There are several disadvantages of multi-agent systems when compared to other power system automation systems.

Management of Large Agent Societies

In a substation containing a large number of items of plant, there is a correspondingly large number of plant agents (the same is true for devices). This creates a difficult task of managing these agents.

Difficulty of Integration with Devices

Although the device agents provide a convenient interface to other agents in the system, the implementation of a device agent is still performed in a similar way to that which would be used for a component of a traditional industrial automation system, and it is necessary to write a specific device agent for each model of device. Therefore, a multi-agent system may not represent a significant improvement in this area.

Inflexibility of User Interface

The major modifiability problems of the architecture that have been identified concern the user interface. It is not possible to automatically modify the user interface for a substation if the layout of the substation changes or new plant or data acquisition devices are added. Also, if new software services are added, the user interface and its agent must be modified to make use of these services.

It might be possible to minimise these problems in two ways: developing a method of automatically generating the one-line diagram and user interface for a substation from a logical description of that substation, and providing a more modular user interface to which new services could be added as they became available. The first of these might be similar to that developed by Qui and Gooi [4], who were able to generate one-line diagrams from a model of a substation. The personal agents discussed in Section 6.2.2 would be used as a powerful and flexible means to promote the design and development of user interfaces.

However, further work might be necessary to provide a means to automatically generate or modify the other elements of the user interface such as the menus, dialogue boxes *etc.*

Possible Performance Problems

The use of ACL messages may degrade the performance of the multi-agent system due to parsing and message passing overheads.

7.8 Summary

This chapter has presented an evaluation demonstrating both advantages and disadvantages to the use of a multi-agent system for power system automation. The major advantages of the multi-agent architecture derive from the use of directories. Because agents locate each other at runtime using a directory service based only on their capabilities, it is possible to add and replace components during the operation of the system. This means that, for example, the user interface agent is capable of obtaining data either from a database or from a plant agent without modification. Another important facet of the multi-agent system is the standard agent communication language, whose defined semantics permit the integration of data from multiple sources. It is also considered that agent autonomy will prove useful in implementing distributed control schemes. However, these are not covered in the prototype. The main disadvantages arise from the complexity of the multi-agent implementation, which consists of a large number of agents and can prove difficult to administer. There are also performance problems, both specific to the implementation described here and generic problems relating to, for example, agent communication. Also, the use of a distributed system can result in increased network traffic and communications overhead compared to a centralised implementation.

Further work is required to fully complete this evaluation. First, it is necessary to install the system in a substation or substations in order to perform a user-focused evaluation involving substation engineers. From the evaluation performed, it is also unclear how the reliability of a multi-agent system compares to that of a traditional system. It would be necessary to examine this question more fully once reliability mechanisms such as redundancy and fault recovery have been developed and incorporated into the prototype.

References

1. The National Grid Company plc (2003) Seven year statement. `http://www.nationalgrid.com/uk/library/documents/sys_03/default.asp`
2. McDonald JD (2003) Substation automation: IED integration and availability of information. IEEE Power and Energy Magazine **1** (2):22–31
3. Lohmann V (2001) New strategies for substation control, protection and access to information. Proceedings of the Sixteenth International Conference and Exhibition on Electricity Distribution (CIRED 2001), IEE Publishing **3**:211–215
4. Qui B, Gooi HB (2000) Web-based SCADA display systems for access via internet. IEEE Transactions on Power Systems **15** (2):681–686
5. Bricker S, Gonen T, Rubin L (2001) Substation automation technologies and advantages. IEEE Computer Applications in Power **14** (3):31–37
6. Hughes JV, Fitch JE, Silversides RW (2001) Substation information project-field experience with internet technologies. Proceedings of the Seventh International Conference on Developments in Power System Protection, Amsterdam, Netherlands
7. Jennings NR (2001) An agent-based approach for building complex software systems. Communications of the ACM **44** (4):35–41
8. Ferber J (1999) Multi-Agent Systems: An Introduction to Distributed Artificial Intelligence. Addison-Wesley, Harlow, England
9. Considine DM, Considine GP (1986) Standard Handbook of Industrial Automation. Chapman and Hall, New York
10. Bernard JW (1989) CIM in the process industries. Instrument Society of America, Research Triangle Park (N.C.)
11. Williams TJ (1989) A Reference Model for Computer Integrated Manufacturing. Instrument Society of America, Research Triangle Park (N.C.)
12. Weedy BM, Cory BJ (1998) Electric Power Systems. Wiley, Chichester
13. Preiss O, Wegmann A (2001) Towards a composition model problem based on IEC61850. Proceedings of the 4th ICSE Workshop on Component-based Software Engineering, Toronto, Canada. Available online: `http://www.sei.cmu.edu/pacc/CBSE4_papers/PreissWegmann-CBSE4-4.pdf`
14. Zhao Q, In H, Wu X, Huang G (2000) Transforming legacy energy management system (EMS) modules into reuseable components: A case study. Proceedings of IEEE International Computer Software and Applications Conference (COMPSAC 2000), Taipei, Taiwan, IEEE Computer Society Press 105–110

15. Evans JW (1989) Energy management system survey of architectures. IEEE Computer Applications in Power **2** (1):11–16
16. Humphreys S (1998) Substation automation systems in review. IEEE Computer Applications in Power, **7** (2):24–30
17. Caird K (1997) Integrating substation automation. IEEE Spectrum **34** (8):64–69
18. Woodward D, Tao D (2000) Comparing throughput of substation networks. Technical report, Schweitzer Engineering Laboratories Inc., Pullmann, WA, USA. Available online: http://www.selinc.com/techpprs/6116.pdf
19. Adamiak M, Premerlani W (1999) The role of utility communications in a deregulated environment. Proceedings of the Hawaii International Conference on System Sciences, Maui, HI, USA
20. Skeie T, Johannessen S, Brunner C (2002) Ethernet in substation automation. IEEE Control Systems Magazine **22** (3):43–51
21. Henning M, Vinoski S (1999) Advanced CORBA Programming with C++. Addison Wesley, Reading, Massachusetts
22. Orfali R, Harkey D (1998) Client/Server Programming with Java and Corba. 2nd edition, Wiley Computer Publishing, New York, USA
23. Somerville I (2001) Software Engineering. 6th edition, Addison-Wesley, Harlow, England
24. International Electrotechnical Commission (2002) Communication networks and systems in substations, IEC Standard 61850, International Electrotechnical Commission, Geneva
25. Clinard K (1999) GOMSFE (generic object models for substation and feeder equipment) models of multifunctional microprocessor relays. Power Engineering Society Summer Meeting **1**:36–38
26. Brunello G, Smith R, Campbell CB (2001) An application of a protective relaying scheme over ethernet LAN/WAN. Transmission and Distribution Conference and Exposition (IEEE/PES) **1**:522–526
27. DNP Users Group. DNP users group website: http://www.dnp.org
28. IEEE Computer Applications in Power Tutorial (2001) Fundamentals of utilities communication architecture. IEEE Computer Applications in Power **14** (3):15–21
29. Woolridge M, Jennings N (1995) Intelligent agents: Theory and practice. The Knowledge Engineering Review **10** (2):115–152
30. Das S, Shuster K, Wu C (2002) ACQUIRE: Agent-based Complex Query and Information Retrieval Engine. Proceedings of the first International Joint Conference on Autonomous Agents and Multi-Agent Systems (AAMAS'02), Bologna, Italy.
31. Sycara K, Paolucci M, Velsen M, Giampapa J (2003) The RETSINA MAS infrastructure. Autonomous Agents and Multi-Agent Systems **7** (1-2):29–48
32. Rao AS, Georgeff MP (1995) BDI agents: From theory to practice. Proceedings of the International Conference on Multi-Agent Systems (ICMAS), 312–319, San Francisco, USA
33. Ingrand FF, Georgeff MP, Rao AS (1992) An architecture for real-time reasoning and system control. IEEE Expert **7** (6):34–44
34. Singh MP, Rao AS, Georgeff MP (1999) Formal Methods in DAI: Logic-Based Representation and Reasoning. In: Weiss G(Eds) Multiagent Systems: A Modern Approach to Distributed Artificial Intelligence. 331–376. the MIT Press, Cambridge, Massachusetts

35. Cohen P, Levesque H (1990) Intention is choice with commitment. Artificial Intelligence **42**:213–261

36. Foundation for Intelligent Physical Agents (2000) FIPA SL content language specification. http://www.fipa.org/specs/fipa00008/

37. Brooks RA (1986) A robust layered control system for a mobile robot. IEEE Journal of Robotics and Automation **2** (1):14–23

38. Wooldridge M (1999) Intelligent agents. In: Weiss G(Eds) Multiagent Systems: A Modern Approach to Distributed Artificial Intelligence. 27–77. The MIT Press, Cambridge, Massachusetts

39. Kaelbling LP, Littman ML, Moore AW (1996) Reinforcement learning: A survey. Journal of Artificial Intelligence Research **4**:237–285

40. Koza JR, Bennett III FH, Andre D, Keane MA (1999) Genetic programming: Biologically inspired computation that creatively solves non-trivial problems. In: Landweber L, Winfree E, Lipton R, Freeland S(Ed) Proceedings of Discrete Mathematics and Theoretical Computer Science (DIMACS) Workshop on Evolution as Computation. 15–44, Princeton University, Springer-Verlag

41. Muggleton S, Raedt L (1994) Inductive logic programming: Theory and methods. Journal of Logic Programming **19/20**:629–679

42. Heinze C, Goss S, Lloyd I, Pearce A (1999) Plan recognition in military simulation: Incorporating machine learning with intelligent agents. Proceedings of IJCAI-99 Workshop on Team Behaviour and Plan Recognition 53–64, Stockholm, UK.

43. Olivia C, Chang CF, Enguix CF, Ghose AK (1999) Case-based BDI agents: an effective approach for intelligent search on the world wide web. AAAI Spring Symposium on Intelligent Agents, Stanford University, USA

44. Maes P (1997) Agents that reduce work and information overload. In: Bradshaw JM(Eds) Software Agents. Chapter 8. The MIT Press, Cambridge, Massachusetts

45. Boyan JA, Littman ML (1993) Packet routing in dynamically changing networks: A reinforcement learning approach. In: Cowan JD, Tesauro G, Alspector J(Ed) Advances in Neural Information Processing Systems. 671–678. The MIT Press, Cambridge, Massachusetts

46. Stone P, Veloso M (2000) Multiagent systems: A survey from a machine learning perspective. Autonomous Robotics **8** (3)

47. Ferguson IA (1992) TouringMachines: Autonomous agents with attitudes. IEEE Computer **25** (5):51–55

48. Fischer K, Muller JP, Pischel M (1995) A pragmatic BDI architecture. Proceedings of ATAL 95, number LNAI 1037 in Lecture Notes in Artificial Intelligence, 203–218 Springer Verlag

49. Foundation for Intelligent Physical Agents. FIPA homepage. http://www.fipa.org

50. Foundation for Intelligent Physical Agents (2000) FIPA agent standard specification. http://www.fipa.org/repository/standardspecs.html

51. Milojicic D *et. al,* (1998) MASIF: The OMG mobile agent system interoperability facility. In Proceedings of the 2nd International Workshop on Mobile Agents, Lecture Notes in Computer Science, **1477** 50–67, Springer-Verlag, Berlin, Germany

52. Fuggetta A, Picco GP, Vigna G (1998) Understanding code mobility. IEEE Transactions on Software Engineering **24** (5):342–361

53. Glitho RH, Olougouna E, Pierre S (2002) Mobile agents and their use for information retrieval: A brief overview and an elaborate case study. IEEE Network **16** (1):34–41
54. Gray RS, Cybenko G, Kotz D, Peterson RA, Rus D (2002). D'Agents: Applications and performance of a mobile-agent system. Software— Practice and Experience **32** (6):543–573
55. Baldi M, Picco GP (1998) Evaluating the tradeoffs of mobile code design paradigms in network management applications. Proceedings of the 20th International Conference on Software Engineering, IEEE Computer Society 146–155 Washington, DC, USA
56. Schwehm E (1997) A performance model for mobile agent systems. Proceedings on Parallel and Distributed Processing Techniques and Applications, Las Vegas, Nevada, USA **2**:1132–1140
57. Brewington B, Gray R, Moizumi K, Kotz D, Cybenko D, Rus R (1999) Mobile agents for distributed information retrieval. In: Klusch M(Eds) Intelligent Information Agents, Springer-Verlag, Chapter 15 355–395
58. Harrison CG, Chess DM, Kershenbaum A (1996) Mobile agents: Are they a good idea. Technical report, IBM T. J. Watson Research Center
59. Milojicic D (1999) Trend wars: Mobile agent applications. IEEE Concurrency 80–90
60. Jansen W, Karygiannis T (1999) Mobile agent security. Special Publication 800-19, National Institute of Standards and Technology
61. Chess D (1998) Security issues in mobile code systems. Mobile Agents and Security, number 1419 in Lecture Notes in Computer Science 159–187
62. Shehory O (1998) Architectural properties of multi-agent systems. Technical Report CMU-RI-TR-98-28, The Robotics Institute, Carnegie Mellon University, Pittsburgh, Pennsylvania 15213
63. Martin D, Cheyer A, Moran D (1999) The Open Agent Architecture: a framework for building distributed software systems. Applied Artificial Intelligence **13** (1/2):91–128
64. Rao AS (1996) AgentSpeak(L): BDI agents speak out in a logical computable language. In Proceedings of Modelling Autonomous Agents in a Multi-Agent World, number 1038 in LNAI, 42C55. Springer Verlag
65. Bordini RH, Hübner JF, *et al* (2004) Jason: A Java-based agentSpeak interpreter used with saci for multi-agent distribution over the net, manual, first release edition. http://jason.sourceforge.net/
66. Hübner JF (2003) SACI - Simple Agent Communication Infrastructure. Programming Manual. http://www.lti.pcs.usp.br/saci/
67. Bordini RF, Moreira AF (2004) Proving BDI properties of agent-oriented programming languages: The asymmetry thesis principles in AgentSpeak(L). Annals of Mathematics and Artificial Intelligence, **42**(1–3)
68. Moreira AF, Vieira R, Bordini RH (2003) Extending the operational semantics of a BDI agent-oriented programming language for introducing speech-act based communication. Procceedings of Declarative Agent Languages and Technologies (DALT). Melbourne, Austrilia
69. Bordini RH, Dastani M, Dix J, El Fallah Seghrouchni A (eds). (2005) Multi-Agent Programming: Languages, Platforms and Applications. Number 15 in Multiagent Systems, Artificial Societies, and Simulated Organizations. Springer

70. Dastani M, van Riemsdijk B, Dignum F, Meyer JJ (2003) A Programming Language for Cognitive Agents: Goal Directed 3APL. Proceedings of the Second International Joint Conference on Autonomous Agents and Multi-Agent Systems (AAMAS'03) Melbourne, Australia
71. Bellifemine F, Poggi A, Rimassa G (2001) Developing multi-agent systems with a FIPA-compliant agent framework. Software: Practice and Experience **31** (2):103–128
72. Pokahr A, Braubach L, Lamersdorf W (2003) Jadex: Implementing a BDI-Infrastructure for JADE Agents. EXP - in search of innovation, **3** (3):76–85
73. Protégé Home Page, http://protege.stanford.edu/
74. Busetta P, Ronnquist R, Hodgson A, Lucas A (1999) JACK intelligent agents - components for intelligent agents in Java. AgentLink News Letter, vol. 2
75. Howden N, Ronnquist R, Hodgson A, Lucas A (2001) JACK Intelligent Agents: Summary of an agent infrastructure. In Wagner T, Rana O (eds) The 5th International Conference on Autonomous Agents, Workshop on Infrastructure for Agents, MAS and Scalable MAS, 251–257 Montreal, Canada
76. Smith RG, Davis R (1981) Frameworks for cooperation in distributed problem solving. IEEE Transactions on Systems, Man and Cybernetics SMC-**11** (1):61–69
77. Foundation for Intelligent Physical Agents (2000) FIPA agent management specification. http://www.fipa.org/specs/fipa00023/
78. Decker K, Sycara K, Williamson M (1997) Middle-agents for the internet. Proceedings of the 15th International Joint Conference on Artificial Intelligence, Nagoya, Japan
79. Kumar S, Cohen PR, Levesque HJ (2000) The adaptive agent architecture: Achieving fault-tolerance using persistent broker teams. Proceedings of the Fourth International Conference on Multi-agent Systems 159–166 Boston, MA, USA
80. Sycara K, Decker K, Pannu A, Williamson M, Zeng DJ (1996) Distributed intelligent agents. IEEE Expert **11** (6):36–46
81. Huhns MN, Singh MP (1998) All agents are not created equal. IEEE Internet Computing **2** (3):94–96
82. Genesereth MR, Keller AM, Duschka OM (1997) Infomaster: an information integration system. Proceedings of 1997 Association for Computing Machinery Special Interest Group on Management of Data Conference 539–542 Tucson, USA
83. Finin T, Fritzson R, Mckay D, Mcentire R (1994) KQML as an agent communication language. Proceedings of the Third International Conference on Information and Knowledge Management, ACM Press
84. Jennings NR, Cockburn D (1996) ARCHON: A distributed artificial intelligence system for industrial applications. In: O'Hare GMP, Jennings, NR(Ed) Foundations of Distributed Artificial Intelligence, 319–344, Wiley
85. The PABADIS Consortium (2002) Pabadis white paper http://www.pabadis.org/downloads/pabadis_white_paper.pdf
86. Deter S, Blume R, Eckehardt K (2002) Generic machine representation in the PABADIS community. Proceedings of e-2002 Conference, Prague http://www.mathematik.uni-marburg.de/~pabadis/publics/e2002.doc.
87. Schild K (2000) Self-organizing manufacturing control: An industrial application of agent technology. Proceedings of the Fourth International Conference on Multi-Agent Systems 87–94 Boston, MA, USA

88. Jennings NR, Bussmann S (2003) Agent-based control systems: Why are they suited to engineering complex systems. IEEE Control Systems Magazine **23** (3):61–73

89. Leitao P, Restivo F (2001) An agile and cooperative architecture for distributed manufacturing systems. Proceedings of the IASTED International Conference on Robotics and Manufacturing, 188–193, Cancun, Mexico

90. Mangina EE, Mcarthur SDJ, Mcdonald JR (2001) Commas condition monitoring multi-agent system. Autonomous Agents and Multi-agent Systems **4** (3):279–282

91. International Electrotechnical Commission (2002) Communication networks and systems in substations. IEC Standard 61850, International Electrotechnical Commission, Geneva

92. Lucas C, Zia MA, Shirazi MRA, Alishahi A (2001) Development of a multiagent information management system for Iran power industry: A case study. In Proceedings of 2001 IEEE Porto Power Tech Conference, Porto, Portugal

93. Downes J, Goody J, Walker K, Baker B, Cooper D (1998) A strategy for substation information, control and protection. Technical Report TR(E)312, National Grid Company

94. Neumann G,Zdun U (2000) High-level design and architecture of an HTTP-based infrastructure for web applications. World Wide Web **3** (1):13–26

95. Foundation for Intelligent Physical Agents (2000) FIPA ACL Message Structure Specification, `http://www.fipa.org`

96. Foundation for Intelligent Physical Agents (2000) FIPA communicative act library specification, `http://www.fipa.org/specs/fipa00037/`

97. Buse DP, Sun P, Wu QH, Fitch J (2003) Agent-based Substation Automation. IEEE Power and Energy Magazine **1** (2):50–56

98. Wu QH, Buse DP, Sun P, Fitch J (2003) An architecture for e-Automation. The IEE Computing and Control Engineering Journal **14** (1):38–43

99. Wu QH, Buse DP, Feng JQ, Sun P, Fitch J (2005) e-Automation, an architecture for distributed industrial automation systems. International Journal of Automation and Computing **1** (1):17–25

100. Corera JM, Laresgoiti I, Jennings NR (1996) Using Archon. part 2: Electricity transportation management. IEEE Expert **11** (6):71–79

101. Nagata T, Nakayama H, Utatani M, Sasaki H (2002) A multi-agent approach to power system normal state operations. In 2002 IEEE Power Engineering Society Summer Meeting, **3**:1582–1586

102. Vishwanathan V, Mccalley J, Honavar V (2001) A multiagent system infrastructure and negotiation framework for electric power systems. Proceedings of 2001 IEEE Porto Power Tech Conference, **1** Porto, Portugal

103. Vishwanathan V, Gangula V, Mccalley J, Honavar V (2000) A multiagent systems approach for managing dynamic information and decisions in competitive electric power systems. In NAPS 2000, Waterloo, Canada

104. Gustavsson R (1999) Agents with power. Communications of the ACM **42** (3):41–47

105. Mangina EE, Mcarthur SDJ, Mcdonald JR (2001) Reasoning with modal logic for power plant condition monitoring. IEEE Power Engineering Review **21** (7):58–59

106. Parunak HVD (1999) Industrial and Practical Applications of DAI. In Weiss G (Eds) Multiagent Systems: A Modern Approach to Distributed Artificial Intelligence. Chapter 9, 377–421. The MIT Press, Cambridge, Massachusetts

107. Shen W, Norrie DH (1999) Agent-based systems for intelligent manufacturing: A state-of-the-art survey. Knowledge and Information Systems **1** (2):129–156

108. Martin D, Oohama H, Moran D, Cheyer A (1997) Information brokering in an agent architecture. Proceedings of the Second International Conference on the Practical Application of Intelligent Agents and Multi-Agent Technology, Blackpool, UK, 467–486

109. Foundation for Intelligent Physical Agents. FIPA Ontology Service Specification, 2001, http://www.fipa.org/specs/fipa00086/

110. Labrou Y, Finin T, Peng Y (1999) Agent communication languages: The current landscape. IEEE Intelligent Systems **14** (2):45–52

111. Foundation for Intelligent Physical Agents. FIPA interaction protocol library specification, 2001, http://www.fipa.org/specs/fipa00025/

112. Buse DP, Sun P, Wu QH, Baker B (2001) An agent based architecture for substation data integration. Proceedings of CIGRE International Conference on Power Systems, Wuhan, China, 551–554

113. Parunak HVD (1997) Go to the ant: Engineering principles from natural multi-agent systems. Annals of Operations Research **75**:69–101

114. Feng JQ, Tang WH, Wu QH, Fitch J (2004) An ontology for knowledge representation in power systems. Proceedings of UKACC Control, University of Bath, UK, **35**:1–5

115. Ma C, Feng JQ, Yang Z, Wu QH (2005) Agent-based personal article citation assistant. Proceedings of The 2005 IEEE International Joint Conference on Web Intelligence and Intelligent Agent Technology, Compiegne University of Technology, France, 702–705

116. The National Grid Company. Specification for substation control systems. NGTS 2.7, The National Grid Company, National Grid House, Kirby Corner Road, Coventry CV4 8JY, April 1998.

117. Russell S, Norvig P (1995) Artificial Intelligence: A Modern Approach. Prentice Hall Series in Artificial Intelligence. Prentice Hall, Upper Saddle River, NJ, USA

118. Hoek WVD, Wooldridge M (2003) Towards a logic of rational agency. Logic Journal of the IGPL **11** (2):135–159

119. Allen JF (1991) Time and time again: the many ways to represent time. International Journal of Intelligent Systems **6**:341–355

120. Shoham Y (1987) Temporal logics in AI: Semantical and ontological considerations. Artificial Intelligence **33** (1):89–104

121. Kowalski R, Sergot M (1986) A logic-based calculus of events. New Generation Computing **4** (1):67–95

122. Tengdin JT (2000) Development of an IEEE standard for integrated substation automation commmunication: specifications for substation applications and key communication performance drivers. Power Engineering Society Summer Meeting, IEEE **1**:136–137

123. Martin KE, et al. (1998) IEEE standard for synchrophasors for power systems. IEEE Transactions on Power Delivery **13** (1):73–77

124. Dublin Core Metadata Initiative. The Dublin Core Element Set version 1.1. http://www.dublincore.org/documents/dces/, 2003

125. Feng JQ, Sun P, Tang WH, Buse DP, Wu QH, Richardson Z, Fitch J (2002) Implementation of a power transformer temperature monitoring system. Proceeding of 2002 International Conference on Power System Technology, IEEE **3**:1980–1983

126. Guarino N (1998) Formal ontology and information systems. Proceedings of the 1st International Conference on Formal Ontologies in Information Systems, Trento, Italy, IOS Press, 3–15

127. Foundation for Intelligent Physical Agents. (2001) Publically available implementations of FIPA specifications. `http://www.fipa.org/resources/livesystems.html`.

128. Chen HL (1999) Developing a dynamic distributed intelligent agent framework based on the Jini architecture. MSc thesis, University of Maryland, Baltimore County

129. Edwards WK (2001) Core Jini. Prentice Hall PTR, Upper Saddle River, NJ 07458

130. Bergenti F, Poggi A (2002) LEAP: A FIPA platform for handheld and mobile devices. Intelligent agents VIII, LNAI 2333, Springer-Verlag, Berlin Heidelberg, 436–445

131. Paton NW, Oscar D (1999) Active database systems. ACM Computing Surveys (CSUR) **31** (1):63–103

132. Gupta G, Pontelli E, Ali KAM, Carlsson M, Hermenegildo MV (2001) Parallel execution of prolog programs: a survey. ACM Transactions on Programming Languages and Systems (TOPLAS) **23** (4):472–602

133. Petrie CJ (1996) Agent-based engineering, the web, and intelligence. IEEE Expert **11** (6):24–29

134. Salton G, Buckley C (1988) Term-weighting approaches in automatic text retrieval. Information Processing and Management **24** (5):513–523

135. Kowalski G (1997) Information Retrieval Systems: Theory and Implementation. Kluwer Academic Publishers, Boston, MA, USA

136. Kretser O, Moffat A, Shimmin T, Zobel J (1998) Methodologies for distributed information retrieval. Proceedings of the Eighteenth International Conference on Distributed Computing Systems, Amsterdam, The Netherlands, 66–73

137. Gray RS, *et al* (2001) Mobile-agent versus client/server performance: Scalability in an information-retrieval task. Mobile Agents: 5th International Conference, MA 2001, Atlanta, Georgia, number 2240 in Lecture Notes in Computer Science, 229–243

138. Buse DP, Wu QH (2004) Mobile agents for remote control of distributed systems. IEEE Trans on Industrial Electronics, Special Issue on Distributed Network-based Control Systems and Applications **51** (6):1142–1149

139. Breg F, Polychronopoulos CD (2001) Java virtual machine support for object serialization. ISCOPE Conference on ACM 2001 Java Grande, ACM Press, 173–180

140. Buse DP, Feng JQ, Wu QH (2003) Mobile agents for data analysis in industrial automation systems. Proceedings of The 2003 IEEE/WIC International Conference on Intelligent Agent Technology, Halifax, Canada, 60-66

141. Baek JW, Yeo JH, Kim GT, Yeom HY (2001) Cost effective mobile agent planning for distributed information retrieval. International Conference on Distributed Computing Systems(ICDCS), Mesa, Arizona, USA, 65–72

142. Moizumi K, Cybenko G (2001) The traveling agent problem. Mathematics of Control, Signals and Systems **14** (3):213–232

143. Xie R, Rus D, Stein C (2001) Scheduling multi-task agents. In Mobile Agents: 5th International Conference, number 2240 in Lecture Notes in Computer Science, Springer-Verlag

144. Rizzo L (1997) Dummynet: a simple approach to the evaluation of network protocols. ACM SIGCOMM Computer Communication Review **27** (1):31–41

145. Johansen D (1998) Mobile agent applicability. Proceedings of Mobile Agents (MA) 1998, Springer Verlag, number 1477 in Lecture Notes in Computer Science, 80–98

146. Tsukui R, Beaumont P, Tanaka T, Sekiguchi K (2001) Power system protection and control using intranet technology. IEE Power Engineering Journal, 249–255

147. Jennings NR (1995) Controlling cooperative problem solving in industrial multi-agent systems using joint intentions. Artificial Intelligence **75** (2):195–240

148. Marriott K, Stuckey PJ (1998) Programming with constraints: an introduction. MIT Press, Cambridge, Mass

149. Yokoo M, Durfee EH, Ishida T, Kuwabara K (1998) The distributed constraint satisfaction problem: Formalization and algorithms. IEEE Transactions on Knowledge and Data Engineering **10** (5):673–685

150. Campadello S, Helin H, Koskimies O, Raatikainen K (2000) Wireless Java RMI. Proceedings of the 4th International Conference on Enterprise Distributed Object Computing, Makuhari, Japan, IEEE Computer Society, 114–123

151. Buse DP, Sun P, Wu QH, Fitch J (2003) Mobile agents for industrial information management, monitoring and control. Proceedings of Intelligent Agents, Web Technologies and Internet Commerce (IAWTIC 2003), Vienna, Austria

152. Delamaro M, Picco GP (2002) Mobile code in .NET: A porting experience. Mobile Agents: 6th International Conference MA 2002, number 2535 in Lecture Notes in Computer Science

153. Wu QH, Feng JQ, Tang WH, Fitch J (2005) Multi-agent based substation automation systems. IEEE PES 2005 General Meeting, San Francisco, California, USA

154. Buse DP, Sun P, Wu QH, Fitch J (2003) e-Automation – integration of information management, condition monitoring and real-time control. Proceedings of International Conference on Intelligent Agents, Web Technologies and Internet Commerce, Vienna, Austria, 470-481

155. Feng JQ, Smith JS, Wu QH, Fitch J (2005) Condition assessment of power system apparatuses using ontology systems. Procceeding of 2005 IEEE/PES Transmission and Distribution Conference and Exhibition: Asia and Pacific, Dalian, China, F0350, 1-6

156. Nichols SJV (2003) XML raises concerns as it gains prominence. IEEE Computer **36** (5):14–16

157. Cahoon B, Mckinley KS (1996) Performance evaluation of a distributed architecture for information retrieval. Proceedings of the 19th Annual International Association for Computing Machinery Special Interest Group on Information Retrieval, (ACM SIGIR) Conference on Research and Development in Information Retrieval, ACM Press 110–118

158. Fuhr N (1999) A decision-theoretic approach to database selection in networked IR. ACM Transactions on Information Systems (TOIS) **17** (3):229–249

159. Kazman R, Klein M, Clements P (2000) ATAM: Method for architecture evaluation. Technical Report CMU/SEI-2000-TR-004, Carnegie Mellon University, USA

160. Szejko S (1999) An exercise in evaluating significance of software quality criteria. ACM SIGCSE Bulletin **31** (3):199

161. Haacke S, Border S, Stevens D, Uluski B (2003) Plan ahead for substation automation. IEEE Computer Applications in Power **1** (2):32–41
162. Fedoruk A, Deters R (2002) Improving fault-tolerance by replicating agents. Proceedings of the First International Joint Conference on Autonomous Agents and Multiagent Systems, ACM Press 737–744
163. Kaminka GA (2000) Execution Monitoring in Multi-agent Systems. PhD thesis, University of Southern California Computer Science
164. Feng JQ, Buse DP, Wu QH, Fitch J (2002) A multi-agent based intelligent monitoring system for power transformers in distributed substations. Proceedings of IEEE/CSEE International Conference on Power System Technology, Kunming, China, **13**:1962-1965
165. Feng JQ, Ma C, Tang WH, Smith JS, Wu QH (2005) A transformer predictive maintenance system based on agent-oriented programming. Proceedings of 2005 IEEE/PES Transmission and Distribution Conference and Exhibition: Asia and Pacific, Dalian, China, F0407:1-6
166. Huxley NH (2001) Intelligent Monitoring of Pre-Term Babies During the First Weeks of Life. PhD thesis, The University of Liverpool

Index

Printed in the United States
66767LVS00002B/280-330